SASQUATCH!

Reports from the Field

ENCOUNTERS ACROSS

NORTH AMERICA

GARY AND WENDY SWANSON

Unless otherwise credited, all images were provided from the files of
Gary Swanson

Cover design by Jim Myers, The Sasquatch Outpost, Bailey, Colorado

Background photo courtesy of sagesolar on pxhere.com:
https://pxhere.com/en/photo/314606

Published by Swanson Literary Group
ISBN: 978-1792920837

Other books by the authors:

They Saw Sasquatch
Sasquatch Encounters
Bigfoot Uncovered
Tracking Sasquatch
Bigfoot Adventures
On the Trail of Sasquatch
Sasquatch is Out There
Squatchin': Study Guide and Field Handbook for Tracking Sasquatch
Hiking Sasquatch Country: Best Hikes In Southern Oregon

Skinwalkers Shapeshifters and Native American Curses
The Last Skinwalker: The Avenging Witch Of The Navajo Nation
We Survived Native American Witches, Curses & Skinwalkers
Skinwalker: Guardian of the Last Portal

CONTENTS

INTRODUCTION

Welcome to the world of Bigfoot/Sasquatch! This is the fifth in our collection of encounters with this mysterious and reclusive creature.

The years we spent hiking the beautiful and rugged Southern Oregon and northernmost California areas brought us into contact with many hikers, campers, gold miners and those rugged individuals who live "off the grid!" While also documenting our hikes with over 1,800 online photos, we met a great many people who mentioned experiences with Sasquatch, and though we had heard of the creature we had never seen one; that is until…that's another story.

Reclusive as these animals are, they do resemble humans in several ways; their appearance is "humanoid" and there is evidence they have family units and a hierarchy of command not unlike humans. They ask nothing from us, and they infrequently take anything from man!

When Doctor Jane Goodall, renowned primatologist and conservationist was asked her opinion about Sasquatch on September 27, 2002, she said, "I tell you that I am sure they exist!"

It all began when many of our readers and friends on Facebook's "Sasquatch Watch" page asked us to include some actual stories of sightings since we lived in an area that Bigfoot seemed to thrive in; so we have!

We put out the word that we would publish actual sightings and encounters and guarantee total anonymity to our submitters (if they so wished), and the stories began coming in.

We included the first submissions in "They Saw Sasquatch," "Sasquatch Encounters," "Bigfoot Uncovered" and "Tracking Sasquatch." There are additional reports by more people who had personal stories of sightings and encounters across the United States and "Tracking Sasquatch" even has encounters from Oak Island, Nova Scotia.

This volume of Sasquatch sightings and encounters spans North America; from British Columbia, to our Eastern and Southern states; Pennsylvania, Maine and North Carolina, and then to the western part of the United States; including Colorado, Montana, Idaho, Oregon, Northern California and Washington state.

Our friends at The Sasquatch Outpost in Bailey, Colorado are featured in this book because not only have they brought us contributors of Colorado Bigfoot sightings, they have been dedicated to researching the Sasquatch and preserving evidence of Bigfoot. Their museum, "The Sasquatch Encounter Discovery Museum" provides a great learning tool for those who have never had a Sasquatch experience or would just like to learn more.

Our many thanks to Jim and Daphne Myers at The Sasquatch Outpost for their support and their dedication to protecting North America's endangered indigenous species of Bigfoot!

Since these mysterious beings have convinced vast numbers of us of their existence, and also of their desire to live peaceably, we have the same goal. Thank you for your courtesy in observing and enjoying the realization that there are some things that man has not yet destroyed!

We have taken great care to guarantee the privacy of every submitter, and due to having proven our confidentiality, more stories have arrived and are in this fifth volume; "Sasquatch! Reports from the Field."

We try to do verification of the stories we agree to accept, and although we cannot guarantee the validity of any submissions, we use due diligence to interview those whose story seems to be outside what the majority have established as the "norm;" if there could be one.

Although our personal experiences with Bigfoot are limited to only a few instances, as we are always accompanied by two curious and noisy dogs; we have interviewed enough people from all walks of life, many with highly respected credentials that we must believe that Sasquatch lives!

INVITATION:

If you have had a personal encounter or sighting of a Sasquatch that you would like to see published in our next book, please send the details and any accompanying photos to:

swanliterary@gmail.com

If your story is published you will receive a copy of the book as our thanks. We look forward to hearing from you!

1 THE SASQUATCH OUTPOST IN BAILEY, COLORADO

Daphne Myers ~ Bailey, Colorado

How it all began….

In May of 2012, two women were taking a stroll in the forest near Bailey. They heard a branch snap behind them and turned to see who or what made the sound, thinking it would be a deer or an elk. What they saw was a large, hairy two-legged creature which paid no attention to them but sprinted past them… and disappeared.

Around this time, my husband Jim and I were working on re-opening the Bailey Country Store, which had fallen into disrepair. While renovating the old building, Jim met one of the women and heard her story. As an avid Bigfoot believer, he was thrilled to think that there were Bigfoot in the area! He had assumed that they only inhabited the northwest region of America. He reasoned that if these women had seen one in this area, this must mean that there were Bigfoot here and other people must have seen them too - they just weren't talking about it.

So he put a map up in the back room and invited people to come and tell about their sightings, marking them on the map. Word got out and the map started filling up. This room became known as The Sasquatch Outpost.

At first I (Daphne) was a skeptic, thinking "How can there be a huge hairy creature out there that avoids detection? It's just not possible!" So whenever someone came in the store wanting to tell their story, I would send them over to Jim. Of course I was listening with open ears, and it did not take many eyewitness accounts to convince me that these people were genuine, they were not making this up, and what they had seen had persuaded them that these creatures exist. You cannot tell someone who has seen a Bigfoot that there is no such thing!

At first all we had in the back room was the map and a few posters talking about Bigfoot. Over the years it has developed into a decent-sized museum, with much information from local researchers, people who have had sightings and photographs of tracks, nests, tree breaks and much more.

We soon learned that when people visited our museum they expected to be able to take home some Bigfoot paraphernalia. We didn't have any! So we began looking on the internet, and found very little available. So there was nothing else for it but to begin designing and printing our own mugs and T shirts. We still sell many of these designs that are unique to our store, but in the meantime the market has become flooded with such items and now we have a plethora of merchandise to choose from. I can safely say that if it exists, we probably have it! From figurines to Lego pieces, bumper stickers, patches, jewelry, keychains, coasters - just to mention a very few!

As the Bigfoot side of our store grew, we realized that the groceries were draining us financially, so after two years of trying to do both, we took the risky step of canning the groceries and focused purely on Bigfoot. We dropped the Country Store out of our name and became The Sasquatch Outpost. The museum we called The Sasquatch Encounter Discovery Museum, and Jim, with his background in graphic design, was able to build a wonderful forest ambiance and a life size model of Bigfoot. It is dark and quiet, just like a forest, and I have literally heard folk scream who were in the museum and suddenly spotted the tall hairy creature peering out at them from behind a tree!

Bailey had become known as somewhat of a Bigfoot hotspot. Following that first encounter, the TV show Finding Bigfoot came and conducted a Town Hall meeting with over a hundred people in attendance, about one tenth of whom had had sightings themselves. We have continued to hold these meetings regularly and more and more people have been emboldened to talk of their experiences once they are sure they are in a safe place, surrounded by fellow believers. We are personally not convinced that there are more sightings in Bailey than in the rest of the Rocky Mountain area; we just believe that due to the presence of our store and museum, there are more people being open and willing to talk of their encounters.

Over the five years that we have been in existence, we have been interviewed by several news stations and been included in the documentary Chasing Bigfoot. Many people who come to our store are curious about this creature, but hesitant to go searching on their own for sightings. So they ask us if we would take them. Jim is always a willing participant in any "squatching activity" so it is not hard to twist his arm! He has several favorite spots where he has had interactions both in the daytime and nighttime, and never tires of trying to get good recordings or camera footage. However, true to nature, the creature continues to be as camera-shy or as wily as his reputation, and it is an ongoing adventure of trying to outsmart the other.

If ever you are traveling in Colorado and can stop by the Sasquatch Outpost on 149 Main St, Bailey 80421 we would love to meet you, and are always ready to "talk Bigfoot". It is our greatest joy to hear other people's stories, and we now have such a wealth of them, if you ever want to hear one, we will happily start pulling them out of the bag! The trick, I have heard, is to get us to stop!!

You can also find us online at sasquatchoutpost.com or call us at our store on 303-816-9383. And of course, if you have a story of a sighting in Colorado, be sure to come by and put your pin in our map and give us another story to add to our collection.

Happy Squatching everyone!

Jim and Daphne Myers

Meet "Frank"

2 WELCOME TO COLORADO

Nancy G. ~ Boulder, Colorado

Ever since I moved to Colorado, several of my friends kept telling me about how they had seen Sasquatch! I had heard about this mysterious sounding animal ever since I was growing up in Oak Park, Illinois.

By the time I moved to Boulder, Colorado, I was already aware of the many stories from around the country about the Bigfoot, so when a couple of my newfound friends began telling me of their personal encounters, I was ready for the game!

I listened one night in a local bar where several of us from the same company were getting to know each other. Since I was the new girl and a stranger to this wild country, everyone wanted me to be aware of the dangers of living in the rugged and untamed mountains. Many of which were so hostile and foreboding, they had never been entirely explored.

I was, as they say, taking everything in with a "grain of salt," when the subject had suddenly been changed to one of the couples encounter with a Sasquatch. It actually sounded plausible, but owing to my Chicago heritage, I couldn't help but be obviously skeptical, and it must have showed, because Demi got up and walked away in a huff.

She soon returned, and I had a chance to explain my natural skepticism, so she got over it and said she understood, but that in this area, stories about sightings and encounters with Bigfoot were so common and so frequent that the locals had learned to listen for telltale signs that would instantly substantiate the truth or falsehood of a storyteller's claim of an encounter.

As the evening wore on, I was entertained by many of the friendly locals who had themselves seen the Bigfoot and some of these people had actually had a close encounter with one.

A man of considerable age had ambled over to sit near our group of circled benches in this large, log-walled corner of the lounge, and after listening to several stories from members of our group, at which he had at times been nodding his head in a sort of affirmation of the accuracy of the stories, but when one of our party stated how private and reclusive the Bigfoot are and how they would never hurt anyone, the old timer suddenly got up and edged into the center of what had become a circle of listeners.

He held up a weathered hand, and said it was time he interrupted our group of "Sasquatch lovers" and tell us something we'd had better know before one of us city slickers got ourselves killed! Then he slipped off his heavy, woolen, red and black checked jacket, and that was when I saw he had only one arm.

The old timer next rolled up the left sleeve of his inside shirt and exposed a scarred stump that showed the part of his arm and the crudely stitched stump that ended just above his elbow! He lifted the bare stub and turned in a circle for all to see. With a slight tremble in his voice, he told us that this is what happens when you forget that Bigfoot is a wild animal!

Then he sat facing us and explained how he had worked on a logging crew that supplied most of the lumber that was used to construct several of the major ski lodges in the area.

His name was "Chigger" Phelps (we never did find out his first name) and he said the construction workers had lived right on the site and seldom had much time off to visit the towns in the area, because they were limited by the short season, so they had to work with very little time off. They were well supplied with all of the luxuries that one could consume, and even had a large and steady supply of what Chigger described as "drinkin' likker." The liquor was forbidden during and even close to a man's shift however, and it was carefully monitored because, as Chigger said, "Booze and chain saws don't mix!"

We were all gathered closely around this very fascinating man as he regaled us with tales of the logging camp life and then he became very reminiscent as his voice lowered and he told of the camp suddenly beginning to have problems with disappearing food supplies, and finally having an encounter with a small Bigfoot; he called it a "monkey monster." He told us how several of the men had surrounded the supply shed when the young "ape kid" was stealing supplies and how they tied up the screeching and biting critter; he said it took about six husky loggers to finally tie it to a large tree near the mess shack.

According to Chigger, nobody made much of the capture, because their group was well aware of Bigfoot, but this was the first time they had ever caught one. He said it was only about five feet tall, but as strong as three men. They offered it food and water, but it just snarled and screeched so loud they finally dragged it into an unused supply shed and put logs up against the door at an angle, and they went back to work. Their intent was to keep it there until the foreman returned from town and let him contact the authorities.

After the first night Chigger said all was quiet. He said he'd taken his watch to guard the shed around 10:00 the next night, and as he returned and was passing the shed, a huge ape came around the corner and slapped the log that was leaning against the door away. He said at the same time, the giant creature grabbed him by the arm, and then reaching out with its other huge paw, it snapped and tore his arm and then physically threw him so hard at the shed that he blacked out!

After pausing to shudder visibly at the memory, Chigger had another drink from his mug of beer and finished by warning us to never underestimate these critters. They're not the tame and friendly beasts some people think they are.

Chigger rolled his sleeve back down, put his coat back on, and as he turned to leave, he explained briefly how the camp doctor was barely able to save his life, and his arm from the elbow down was never found. The sincerity of this old timer made a believer of me!

Rocky Mountain National Park

That was my introduction to the Boulder, Colorado scene. Even after four years here of having done my share of camping and hiking, I still haven't seen a Bigfoot! Nor do I really care; my arm hurts just thinking about it.

3 BIGFOOT AT THE BEAVER POND

Joyce H. ~ Denver, Colorado

I read your book "Bigfoot Uncovered" and was surprised the first story was in Colorado, where I live. I had an incident happen to me when I was in my teens. I am 62 years old now and will never forget it. When I think back on that day, I feel my heart rate go up and feel emotional changes. I wish at the time I would have got a better look, but being scared the flight instinct took over. My hair still rises on my arms when I go over what happened.

Back in the late 60s or early 70s my brothers bought a piece of land in Summit County, Colorado. Because I come from a family of 10 kids, my oldest brother is 20 years older than me, so he and a few other siblings pooled their money together to get the land.

Several times a year, mostly in the summer, our family would all go up and pitch tents and have a lot of fun. I was never really afraid except during the normal campfire stories the grownups told us around the fire.

The outhouse was put about 50 to 100 feet from our camp site, so at night, I remember never being afraid to walk to it; of course there were two or three of us kids going to it at once. Some of us would mention hearing banging, but being kids we would giggle and try to scare one another. But the noises were real.

Footbridge leading to the outhouse ~ photo courtesy of Joyce H.

To get to our cabin you'd take Hwy 9 along the Blue River out of Dillon/Silverthorne. There is a Blue River Campground just across from a forest road. Going up the road, which was all dirt, you'd go past a fork that took you to a place called Rock Creek. Going past that, it is all uphill toward the top of the mountain. Then we'd come to a cattle gate that had to be opened and closed as the owners of the land lived in a cabin there year round and rented little cabins (one roomers) out. They also had cows. They had to let people through as this was a forest road.

The adults in my family found this area while looking for a place to stay on a stormy, rainy night, which eventually led to the land purchase and building of a cabin soon after. We always respected the older couple's land.

About five minutes further up the road past their place was our land. There were two others cabins near ours that were built around the time we put ours up, but hardly anyone ever visited them.

Past our cabin the road went up further till it came to a small open space with willows on the edge of a beaver pond. Until you walked near, you would not even think there was a pond beyond the willows. The road kept going, but became smaller and overgrown. If anyone knew of this place, they parked in a little opening and then started to hike on the trail that was at the end of the road that went to Slate Lake. It was a ways to Slate Lake; not just a there and back day hike, so not many people knew of that location to get to the lake. You could count the people on probably two hands who came to that trail head yearly.

Getting to that point of where the trail started at the end of the road was a nice walk from our cabin. It was less than an hour to get there and back, so all the areas I mention were not that far from our cabin.

On the way to the beaver ponds ~ photo courtesy of Joyce H.

My oldest brother's son (my nephew) and I were really pals despite our age difference of four years. I guess as a teenager I still enjoyed doing tomboy stuff and he was just the one to do it with! He often, on the trips to the cabin, was able to bring his little tote-goat (much like a minibike but built more for rugged riding).

Below our cabin, between it and the outhouse, was a small stream about two feet wide. If you followed the stream up where it was coming from it would wind up around a cabin built up from ours and go close to the road then kind of taper off into a larger opening that was fed by the beaver ponds.

I remember sometimes when we would go up and there would be no water in the stream, so my brothers would go up with shovels and open up the dams the beavers built. South of that pond beyond the willows and the trees was another pond up a bit higher. Then, going up higher was another pond, and then the last one was what we called the "Lilly pond," because it was full of lilies. The Lilly pond fed all the other ponds with whatever water it was getting from wherever. It was a very rugged and remote place at the time, full of thick woods and willow beds.

One late morning or early afternoon it was there at that first opening where the small stream running from the beaver pond became wider before going into the beaver pond itself that we both saw it! My nephew and I were riding up and down the road on the tote-goat, using that open area to go up and turn around and back on the road towards the cabin. It was our first ride of the day.

As we approached the opening and my nephew began to turn, he suddenly said something and went so fast whipping around I was falling off the bike as it really wasn't built for two people to ride comfortably. I was straddling the seat trying to pull myself back up on it fully when I saw it. There squatted down by the edge of the mouth to the stream was a very large, dingy, whitish-gray (more dark than white) something.

It was reaching down into the water and its arms were between its legs, and its knees were to the sides of its face. It turned, but didn't stand completely when our eyes met. Its face was so human, but yet like a primate. Long, thinish hair covered its body. The facial hair was the darkest. The eyes seemed droopy, and when I thought about it, there was a sadness to them. It rose up and turned, and it was then I knew it wasn't just any animal I was seeing, nor was it a human, as it was much bigger than a human and wearing no clothes.

This all happened so fast. I got back on the goat and turned back, but it was gone. We hurried back and told the grownups, but they laughed. Probably because they had scared us with stories of the so called "Side Hill Gouger" and thought we were trying to get back at them. So we let it go.

We were too scared to go anywhere the rest of the day and messed around the cabin talking about what we saw. We left the cabin the day after, as the weekend was over and we headed back to Denver.

I often wonder if this creature was older and somewhat deaf because it didn't seem to hear the tote-goat 'til we were right up on it. We visited the cabin the next few years and somehow lost our fear and began exploring again, but keeping the thought of what we had seen in the back of our minds.

I married and moved out of state. Years later, I divorced and came back to Colorado with my son, and took him to the cabin several times. I told him the story, and he has his own thoughts of the noises he has heard and feelings of being watched as he took his solitary little walks into the forest when he was 10 years old. He thought I told him the story to keep him close to the cabin.

Unfortunately, my nephew died the year my son was born (1980), so this was something I felt so blessed to witness with him, but also sad he isn't here to back the story up.

The family cabin ~ photo by Joyce H.

As I mentioned, a few years after buying the land, the family began to erect a prefab cabin. It had two bedrooms, a kitchen opening into a large family room. The best part of it was the front deck. Because it was built on an incline the front was about 10 feet above the ground. There were no stairs, so the only entrance was the back door which had a step up. As kids we slept on the deck at night and heard many strange noises, but never really tried to guess what they were. But today hearing other peoples' experience with Bigfoot, I realized we were hearing the same type of noises that they heard. It is possible they were made by the Bigfoot we saw that day.

After everyone grew up, the cabin and land were sold, as it was getting more busy with people up around there. The old road has a regular street sign on it now where it turns off Hwy 9 and trees have been taken down and more modern homes have been built where the cabins were (they were all sold and taken down). As I look on Google Earth at the area, I feel bad for the development in the area that was once so wild. Whatever creature lived there before is surely long gone, but I saw it and I am lucky to be a believer. I never will forget.

The framed picture of the trail to the beaver ponds that Joyce has on her wall

Publishers note: We went to Google Earth as Joyce suggested. It appears the beaver ponds are still there in addition to at least two very large homes. Joyce describes the Sasquatch as having been a whitish-gray color; we speculate that this may have been an elderly creature.

4 FISHING WITH SASQUATCH

By Tina A. ~ Colorado

My story starts up by Buena Vista, Colorado during the summer months (July or August). We were staying at some local cabins that have a fishing lake. There were six of us on the back side of the lake; chatting, fishing, and having fun, but not catching any fish.

Buena Vista, Colorado
A.   Atencio [world66.com (https://creativecommons.org/licenses/by-sa/1.0)], via Wikimedia Commons

It was just starting to be dusk and we were thinking the fishing would pick up, so we had started to settle down and get serious about fishing. All of the sudden the 40 foot pine trees above us started to

24

snap and then there was a huge kerplop right in front of us in the lake.

A large heavy rock (from the ripples it was at least a foot and a half wide) was thrown over us, hitting the tree and knocking branches down. We all froze for just a minute, not hearing anything moving or breathing. We grabbed all our stuff and got out of there!

We no longer are on the back side of the lake around dusk when we stay at the cabins.

That's my story.

5   WE SAW SASQUATCH

Anita ~ Castle Rock, Colorado

My husband and I took our two young children on a vacation this last year to Dinosaur National Monument. We live in Castle Rock, Colorado, so for fun and to avoid the Denver traffic we cut across on Highway 67 up through Bailey and on to Highway 70.

It was kind of a slow start, but my sister and her husband had told us about the most unique store in Bailey called "Sasquatch Outpost." Well we found it, and it was just like Julie explained. Our kids just went wild in there like Julie said hers did. This very distinctive store is unlike any other tourist attraction we've ever seen and we travel a lot! They even have a museum dedicated to the Bigfoot animals, and in the store, there were enough items, books, photos, clothing and everything else related to Sasquatch that one could imagine!

By the time we were able to drag the kids and their bundles of souvenirs out, we were pretty sure the Sasquatch had to be a real being. They had so many things devoted to it and so many books written by those who have seen and encountered the animals; we came away with little doubt of its existence. They even had a map of sightings of Sasquatch in the area!

We really enjoyed our trip to the Dinosaur Monument in Utah and even the girls had fun, even though they pretended it was boring. After spending two days in the area, we got really adventurous and returned by way of Steamboat Springs and down Highway 40 to the Eagles Nest Wilderness Area.

Eagles Nest Wilderness Area
Kristin [CC BY 2.0 (https://creativecommons.org/licenses/by/2.0)], via Wikimedia Commons

Just before we got to Highway 70, we had stopped for lunch where our picnic table in the rest area off Highway 40 had a most beautiful view. No other cars were even in the area, so after we ate, we agreed to take the girls on a short hike.

There were signs advising people to watch for bear and stay on the trails, so I didn't have any desire to go too deeply down the path where Abby and Jane had gone bolting way ahead to take advantage of a little exercise, and we were about ready to "whistle" them back when they stopped dead in their tracks!

When Tim and I finally caught up to them, they still hadn't moved; not even a foot! I had never seen them so still and Tim whispered to me, wondering what was wrong. It was as though they were frozen in place. The girls heard us then and Abby slowly raised her hand and barely waved it as a signal to stop. Tim and I stopped, and then inched our way forward until we were almost alongside of the girls, and following where there eyes were staring off in the distance and downhill about two hundred yards, we saw two dark brown shapes that at first I thought were bears. I had never seen Abby and Jane so quiet; they seemed petrified!

Then Jane whispered only one word, "Sasquatch." Tim and I glanced at each other, and then all four of us continued to stare at the two shapes that were on all fours, drinking at a pool beside the plunging creek. With the noise of the cascading water, we could understand why we had been able to get so close without being spotted.

Never in their lives had the kids been this still, and although I was still convinced that we were looking at two young bears, the two animals suddenly rose up on two legs and began to continue downward on the trail. Almost in unison they both turned to the sound of a carful of tourists, with their music blaring on the road above. It was at that moment their gaze turned slightly to their left and all six of us had direct eye contact with nobody moving a muscle.

Pikes Peak Highway sign

Then, somewhere in the distance a dog barked. That was all it took to overcrowd this stage, and the two (now obvious) Sasquatch ran quickly down the trail and within seconds dashed into a heavily forested area on our right.

We saw Sasquatch! All four of us will never forget the excitement of that moment when we realized that America's most secretive creature really exists!

The girls were insistent on returning to tell our story to the folks at the Sasquatch Outpost, but it was too late in the day to make it before they closed and both Tim and I had to be back on our jobs in two days, so we decided to put our adventure into a story for your next book, as we bought several of your current books at Sasquatch Outpost and enjoyed the experiences by all of your submitters. It's nice to know we are not alone! Please accept our submission and say "Hello" to the great folks in Bailey!

6 SARVIS CREEK BIGFOOT

By Dale ~ Colorado Springs, Colorado

Call me Dale. In college, my best friend Charles and I were nicknamed "Chip and Dale," and it stuck. Recently, I had been employed in Colorado's newest, and what we sarcastically refer to as the "recreation industry"; marijuana growing!

Having worked hard and been able to stash away a goodly supply of weed on the side, Chip and I were ready for a well-earned vacation. Not only had we made a lot of money, but we had enough of a pot supply scarfed off to afford a year or two off. When we needed money we could sell some "bud" and have no worries.

To reward ourselves for our cleverness, we bought a shitload of equipment and supplies and headed out for a well-deserved vacation in Colorado's beautiful wilderness in my brand new Jeep. All I need say is that we went into the Sarvis Creek Wilderness area. A lot of it is on private property, but Chip had contacts, so we were able to go into a very secluded and remote area, and after passing through Kremmling and filling the gas tank and two Jerry cans, we headed on trails that I dare not reveal. We stopped long enough to pick up a buddy at the Latigo Guest Ranch before heading into the wilderness.

Eventually; we'll just call him "Buddy," led us to what both Chip and I agreed was the best campsite we ever could have imagined. The creek we had been following wound all over the place, but where Buddy led us, the creek bed sort of circled an upraised mound and it came almost full circle and almost met itself (Buddy said one day there would be an oxbow lake here) and created a very wide, and quite deep channel where you could clearly see trout darting here and there as you walked alongside it.

Leaving Gore Canyon near Kremmling, Colorado
https://www.flickr.com/photos/locosteve/5649625878

About fifty yards from this mound was a beautiful stand of what Buddy said were Ponderosa pine, but being originally from Florida; if it isn't a palm tree, it's a stranger to me. There must have been several hundred trees of varying sizes in the center of a scenic, grassy meadow. All around us the hills sloped upwards to the majestic mountains on three sides, and the almost invisible trail by which we had come up that descended into the bluish canyon.

Almost dead center in this magnificent grove of ancient pines, there was an open area about 40 feet across, and it appeared to have at one time had a log home sitting there. On closer inspection, there were remains of a sort of foundation made of large stones in a kind of square, but many had been moved, and there were remnants of burned and blackened logs. Most of the logs were charred beyond anything but ash made solid by years of rain and snow. On one side, the timbers were visible and one could see plainly the remains of walls cut flat on the inside and left natural on the outer walls.

What a beautiful scene this must have once made! The three of us talked about how we each imagined life would have been back then, and if a person had never known of the luxuries we have today, this must have been grand for its time.

By the time we set up the small safari tent, rebuilt the main fire pit, cut dead chunks of firewood, and organized about as good as we needed for the first night, the sun had gone beyond the tall peaks in the distance. We had enough light from the sky, but the meadows surrounding us were a hazy gray and the forests ranged from gray to coal black. The wind had turned quite cool and we were glad that we had dug out and enlarged the fire pit. Adding a few more large stones to the sides allowed us to use the larger logs we had made short work of with Buddy's chainsaw.

The roaring fire was even more comforting as it soon became totally dark; as even the moon was covered with clouds. There we were in this huge clearing, surrounded by hundred foot pine trees and the whole huge area lit up enough to walk around anywhere within 200 feet; as light as if it were only dusk. You can probably tell how much we were all enjoying ourselves!

The next morning, we rekindled our fire and with the camp stove going, we soon sat down at our small folding table for breakfast. It was then that Buddy asked if we heard the Sasquatch in the night. I hadn't heard anything, because my hobby has always been shooting handguns which took care of my ears, but Chip said he had listened to a couple of woodpeckers as they were tapping out their what he thought were messages. That's when Buddy reminded us that birds roost at night, and what the sounds had been were made by at least two, maybe three, Sasquatch!

That came as a real shock to me, as Chip and I are more of the "city slicker" type. Buddy elaborated somewhat; not to frighten us, but pretty much to say that we must be constantly alert. He said we needn't be scared, but that we had best carry guns everywhere we went; even while in camp just in case.

That night, we all heard the rapping, because after we turned in, a series of raps came from what seemed like two or three different directions, only this time, we all were wide awake listening. We must have discussed it on and off for two hours before finally falling asleep.

The next morning at breakfast, we discovered just how close they had come. If we had happened to step out of the tent at the right time, we may have met face to face! There were tracks that plainly showed in the sandy area that we had cleared from around our fire pit about five feet out to prevent sparks from starting a fire. We carefully circled the area around us, and altogether, we counted eleven foot prints, that except for their huge size could have been human!

We could see them clearly enough to know they were bare feet; five toes, but although the toes were prominent, it looked like there were claw marks in front, but we couldn't be absolutely certain. It was more difficult to see more than impressions outside our main camp area, but they had all three circled the Jeep many times. We decided to take inventory and it didn't appear that anything was missing; until later when we decided to go for a hike toward the nearby creek, we discovered two of our hiking staffs were gone. Buddy's was there leaning against the tent, but Chip and I had left ours leaning against a large pine tree about 20 feet to the side of camp. Those we will miss, because they were each six feet tall and we had bought them at a roadside outdoor shop only a month before. They were carved with Native American symbols and very ornate. This establishes that the Sasquatch have good taste!

Over the next week, we experimented a bit by leaving the occasional piece of food on our table at night, but nothing was taken; although we did find a few fresh Sasquatch foot prints, but they avoided coming so close again, and it took us a few days to figure out why. We had taken to sweeping the area around our tent and camp fire area with our small camp broom in order that we might examine the area for fresh tracks each morning. The fact that they stopped walking there gave us to realize that they understood exactly what we were doing it for.

This fact was further evidenced when we very discreetly began sweeping around different areas of the camp under the guise of doing other chores. Then we again saw evidence of their having been in our camp! This cat and mouse game went on for several more days, until the morning we woke up to a low hanging smoke hovering in the air. The mountains in the distance were barely visible and the smell was faint, but definitely from wood or grass. We figured it may have been a local resident from one of the few ranches burning brush, and after the sun was full up, the smoke was gone, so we didn't pay it any more attention.

Late that afternoon, we were all sitting at the place where we had first come into the meadow where the water coursed around this mound, and there in the east we noticed what looked like a thunderstorm forming, but it had a strange look to it, and in a couple of minutes it had shifted almost entirely back behind the mountain. Being as how thunderstorms in this country can be on you sometimes before you even know it, we kept a wary eye out in that direction, but forgot all about it later.

The next day we again returned to the area of the meandering stream with fishing gear, but to no avail. We reasoned that this particular location had a bountiful supply of fish, but the wary trout could see us coming.

As we were sitting up atop the small, central hill, Buddy suddenly pointed toward the mountain and told us to look closely at the base of the tree line; it took a while to concentrate my focus on the fine line between trees and grass, but there it was! A large, dark animal that at first I thought was a bear preparing to climb a tree, but it was actually walking on its hind legs.   That's when Buddy said, "Sasquatch!"

Now, knowing what it was instead of trying to force my brain to see a bear; I could plainly see the large, erect beast walking with a lurching gait that must have covered a yard with every step. Then I alerted my companions that there were two more; slightly shorter ones following close behind. The shimmering air near the ground made the Sasquatch difficult to see, but there was no mistaking the fact that they were really there! Then, as a quick settling of the air quieted a moment, we could clearly see a much smaller one walking on the forested side of the two shorter Bigfoot.

They soon passed behind the ridge and were gone. Coming behind them marched a group of elk, and as we watched, it became a rather large herd moving at a quick and steady pace in that same direction. I was about to joke about a "convention of the animal kingdom" when there suddenly appeared a huge, steadily growing cloud in the same direction as the one we thought was a gathering storm two days before.

Now, as we stood, watching in awe, it grew steadily until it covered the entire eastern sky. Then we were hit with the heavy smell of wood smoke. Forest fire! Even though it may have been over a hundred miles away, it could in reality be only ten miles and coming fast. There was no way to tell, so we beat feet!

This turned out to be the "Silver Creek Forest Fire" near the Roosevelt National Forest. Since it was July of 2018, we knew that no matter what, this was the perfect combination of long term heat and dry terrain that forest fires thrive on. This was why we were so careful to clear extra ground around our camp, because when you are underneath those majestic pines, one spark can start a fire that can destroy a thousand year old forest in days.

Well, we fought waves of smoke on our return, but we kept the headlights on and went as fast as we could back to civilization. As abruptly as the forest fire came up, our lives changed. I've not seen Buddy since and Chip moved to western Montana, while I went to work for a waste disposal company in Grand Junction.

I'll never forget the excitement of finally seeing those Sasquatch I had heard about for so many years! That's why I enjoy hearing about others who have shared my adventure. I do hope the ones we saw were able to safely escape the fire.

Publishers note: According to the U.S. Forest Service, the Silver Creek Fire started on the afternoon of July 19, 2018 and burned 20,120 acres. It was determined that the fire was started by lightning.

7 DISTANT ANCESTORS

David P. ~ Ft. Collins, Colorado

My experience with a Sasquatch came as no surprise to me, as my family has owned our rather remote hunting lodge for a very long time. I won't mention precisely where it is; only that it is in the South Platte River valley. My brother and I grew up with the word "lodge" ingrained in our thinking, however in actuality; our lodge is a very solid and attractive two bedroom log cabin. Not that "cabin" makes it smaller, as we have a large living room with a river stone fireplace that takes three foot logs.

This last year, my brother Tom and I were ready for our annual deer hunt, but a week before, Dad fell and sprained his ankle while out sighting in his ought six. Even in a state of severe pain, Dad insisted that we still go as planned; he just brought a box of reading material. So here we were, keeping up a twelve year tradition.

Tim and I each had our favorite stands in opposite sides of a heavily treed and quite tall hill that started rising about three hundred feet behind the cabin, and rose to a height around six hundred or so feet. This hill became a round-topped ridge that stretched from our property line into a huge area of thickly covered slopes, short valleys, several crisscrossing creeks, and even several small ponds. The ponds were surrounded by cattails and had really soft bottoms. This was ideal country for deer, and normally we filled our three tags the first or second day.

This year was different, and as the three of us sat in front of a toasty fire on the third night, we talked about what had changed. Two full days and neither Tom nor I had spotted any deer! Had we not been what we feel are very experienced hunters, and having hunted in several states, we wouldn't have been so surprised.

So, as we sat around licking our wounds, Dad mentioned that this was not so unusual at all, and in his much greater knowledge of this area and of shared experience from his father, our grandfather, he suggested we slightly alter our hunt on the morrow. Dad advised us to carefully watch the side trails, as our usual method was to be aware of deer tracks, but Dad wanted us to begin stepping off the obvious main routes and occasionally deviate to check out a few of the less traveled paths and most definitely the trails with fresh tracks that looked to be seldom, if ever, used by but a few animals.

The next morning, Tom and I avoided our favorite routes and headed for an area further from our usual spots. Since the three of us most always had three bucks hanging by the second day, we had become spoiled and had seen little reason to "sightsee." Besides that, in this heavily treed area, dotted with meadows and streams, surrounded by thousands of magnificent pines, along with large forests of oak, elm, birch and so much beauty, it is easy to become spoiled by such majesty.

So today seemed like we had a primary mission to analyze the hunting grounds that we had so long taken for granted and we couldn't accept the fact that the deer population was not what we had expected. So where were they? The weather had not changed from past years; northing seemed different from every previous hunt, so why? Tommy and I deviated by an extra half mile to an area that we had never spent more than time to cross through to a resplendent mountain lake two miles beyond.

When we reached a large, round-topped hill about 600 feet high, we agreed to split and go in opposite directions and hunt around it with the knowledge that we would eventually meet, so any shots would need to be at an angle away from the hill enough to not endanger each other. The terrain was similar to where we hunted the days before, and it was about an hour since Tom and I split when I noticed a strange track off to the side of the trail on my right. It seemed as if to be that of another hunter, and there were more tracks all along the grassy area off to the side of the trail.

I thought back to the last two days before Dad expanded our thinking, and wondered if the same tracks may have appeared alongside some of the other animal trails I had walked? My mind dismissed any mystery as I made the assumption that it had been another hunter who had deliberately stepped alongside the trail to hide his prints from the deer, as though they could consciously recognize signs of human presence by their tracks. Dumb thought; I know!

Anyway, I stayed on the more packed surface of the trail, occasionally glancing at the accompanying prints in the grass alongside, until; the grass thinned suddenly at a point where the recent rains had widened a crossing stream to having washed out the grasses and the footprints, as well as the deer racks on the main trail. Then, after crossing this washed out area, I observed a continuation of the deer tracks and off to the right, the human tracks. That's where I stopped dead in my own tracks!

In the still-wet sands, the tracks I had been following for the last miles were perfectly clear now and they were bare footprints. Perfectly imprinted in the damp ground were the largest footprints I had ever seen! Placing my size 12 hunting boot beside one of these prints, heel to heel, the other print was at least another six inches longer and three inches wider than my foot, even with the extended soles of my hunting boots!

The realization that a non-human creature of so much larger proportions had recently walked this same trail made me acutely aware that I may be in danger! My .30-06 carbine didn't feel as powerful in my hands as it had before. I decided to stay where I was until Tom came around to meet me, and I was about to find a place to sit when I heard a gunshot not far ahead. Figuring that my brother may have just succeeded in getting a deer, I reckoned I'd show him my find as I helped him carry the deer back this way, so I trotted off in the direction of the shot, and 10 minutes later, we were carrying a nice, young buck between us, its legs tied to a long branch we held on our shoulders.

My excitement in finding a track must have seemed to Tom to diminish his triumph in such a fine trophy, but once I had seen the large tracks, I had begun to envision what I just knew had to be the elusive Sasquatch, so it was difficult to be too enthused over the deer. I spotted the stick I had shoved into the bank with the piece of orange plastic flagging signaling in the wind, and in my excitement, I almost dropped my end of the deer, as I told Tom to put it down and follow me to my discovery.

We both covered the distance quickly, and as I stopped, gasping for breath, I thrust my hand forward to point at my evidence, but the prints were gone! They had been brushed out by a pine branch that lay directly alongside the spot; it had plainly been used to sweep clean the tracks and it was even more evident, because the culprit had even erased my own recent footprints!

Tom could see my frustration as I just stared in absolute shock, when he said, "Look!" As I raised my line of sight to follow his pointing finger, I could plainly see a large, brownish-gray figure climbing at a face pace up the side of a steep hill. The creature was rapidly going up the steep slope on two feet while using its giant hands to reach out and grab the young forest of saplings to pull itself upward so fast, that in seconds, it had disappeared! We stared after the Sasquatch for several minutes without either of us saying a word.

When we arrived back at the lodge, Dad was standing on his crutches on the porch. He had seen us coming far away, and we remarked that he must have been out there the entire time.

Having tied the deer's hind feet to the lower branch of the large apple tree in the front yard, we burst out with our story of the Sasquatch; figuring Dad would be shocked. Instead, he selected one of the wicker chairs on the deck and motioned us to take seats opposing his; which we did while exchanging curious looks between us. Then after lighting his pipe, Dad told us that that was why he wanted us to widen our tracking perspective and become more aware of our surroundings. He went on to explain that he had noticed this same phenomena in seasons past; as his own father had experienced before him.

The Sasquatch were a nomadic group that occasionally seemed to be forced by extremely cold weather to leave their higher mountain retreats to move down into the surrounding valleys. He went on to say that since our property abutted the main series of valleys directly below their exit route, these creatures were never observed by any of our neighbors, because our land effectively shielded these "big furry guys," as Dad called them.

He told us he'd never mentioned them before to us, because in his words (I'm paraphrasing here), "You clowns would have had every lug-headed, dumbbell friend of yours up here and ruined an entire population of survivors of some long forgotten race of creatures that may even have been our ancestors!" He went on to say that our great-great grandfather may have been one of them! Then with a huge grin, he continued, "Why that one you saw today could very well have been your distant cousin." Strange as it sounds; I wonder? I can still hear Dad laughing!

8   BIGFOOT IN THE LAND OF COVERED BRIDGES

By Tom D. ~ Washington Co., Pennsylvania

I am 47 and have lived in Washington County, Pennsylvania my whole life. My family has had land here for three generations my grandpap's dad came from a small village in Italy to give his family a chance for a better life in America. In total, my family owns over 2,000 acres.

In the fall of 2016 my life turned upside down while building a blind for hunting. I was approached by a huge dark figure on two legs. It was growling and making such a racket I ran!

I came back three days later only to have a deer almost mow me over! Again, I heard the growling and racket, but this time when I ran, it followed. When I looked back, I saw a nine foot, hair covered ape man swaying in trees above me! I never went back.

A month later while rifle hunting on opening morning, a huge female Bigfoot ran out of the trees, looked at me and disappeared into some of the thickest terrain I've ever seen! I know it was a female, because I was looking at her through the 4 x 12 scope on my rifle. The hair on her chest was not as thick as the male Bigfoot I had seen previously, and I could see her breasts moving as she ran.

Publishers note: We received an update from Tom about his Sasquatch neighbors just prior to going to press.

This is what he had to say: "My homestead now has a family group of Bigfoot living here, and I've become very in-tune with them. I posted my property (no trespassing) for the first time in 70 years to give them peace, but, they gotta' tolerate me. So far they've taken few deer and some small game; so far, so good."

9 THE DAY I ALMOST DIED

By Frank M. ~ Bend, Oregon

I had a near death experience a couple of years ago, but until now, my wife Betty was the only person I dared tell. Reason was, I felt I may have been perceived to be out of my mind! Now that I have retired, I no longer care what anyone thinks, so I need to report this incident to maybe help clarify the fact that the Bigfoot creature really lives, and I am proof of that, because if one hadn't come to my aid when it did, I'd have died seven years ago.

My experience took place in the Three Sisters Wilderness area in central Oregon. My hunting partner Hiram and I were elk hunting, and after a tough morning spent climbing, sitting for long periods, and scanning open slopes and glassing further off openings in the heavy timber, we arrived at a place we estimated to be about a mile or two uphill from our pickup. Having hunted these slopes successfully for the past six years, we had a pretty good idea of how to "work the area."

We decided to split up as we had done two years before, and tossed a coin to see who would make the difficult climb. I lost the toss, so Hiram would continue northward along the top ridge which was a gradual descent across the front of the slope to a valley below that wrapped gradually around to an open area about a mile below where we were standing. I on the other hand, would take the more hazardous route on a series of elk and deer trails that sort of zigzagged down through the heavy forests below.

It was our plan that when we both had gone about two hours, that we should, based on past experience, both be at the lightly treed and more open meadow that ran for about fifty miles; like a sort of elevated shelf along the upper part of the valley.

A creek on one of the Three Sisters, which are volcanoes in the Cascade Range in the Three Sisters Wilderness in central Oregon
U.S. Forest Service [Public domain], via Wikimedia Commons

We had hunted this way over the years, and were extremely cautious when taking a shot, because we would end up circling and working directly toward one another. That's where our trust in each other to never take a shot without a ground barrier, tree, or rock being behind our line of sight and the game animal, because our partner would be behind that!

It was a beautifully warm fall day, and I made good time on the unseasonably dry trails and rocky ledges that could be so treacherous at times. Taking time to periodically stop to scan the slopes and carefully listen, I had gone perhaps a mile downward when out of the corner of my eye I caught a movement on my left. There; about 300 feet from me was a large, almost perfectly flat meadow. It was a plateau that stretched back into the side of the slope, and about the size of a baseball diamond. I could hear the splashing of a noisy creek as it cascaded down the towering cliff above on the left, and at the same time, I caught a glimpse of a grayish-brown patch between two towering pine trees in the dark forest beyond. I was almost shaking in my excitement, and my mind was already gloating at how easy it was going to be for me to tie a rope on my elk and bring it easily down the slope to surprise Hiram with my success.

As I slowly allowed my feet to move forward, my eyes were busy scanning every place before my feet touched the ground, while at the same time, continually glancing up to keep a wary eye on that patch of hair behind the trees that I was now certain was a prize, bull elk!

It's funny how the human mind can travel through so many subjects simultaneously; on one hand I was taking my shot with shaking crosshairs, while the next moment I was accepting the Boone and Crockett trophy, while the next, I was hanging the trophy rack on the wall of my local lodge. The part I hadn't envisioned entered my mind with a crashing crescendo of frightening reality; I was falling!

I had been vaguely aware that the stream I had been coming to was dropping into more than a rocky area, it was falling into a very deep pond of beautifully clear, cold, azure blue water. As I made my stalk toward my trophy, my subconscious mind had been aware enough to keep me away from the edge, but had I dared to divert my eyes even for a second, I may have noticed the very thick growth of blackberry bushes that had now entangled themselves around my hunting boots. I knew without looking, what was happening, and I had taken one more forceful step to try extricating my entangled left leg. I did not dare to even risk looking down to assess the situation for fear of making any movement to alert my trophy.

That was my "one step too many," as it seemed as though the entire blackberry bush had come alive as a carnivore. Now, I was not only still falling, but I was falling to my left! My rifle had departed my grasp in that careless manner that my mind immediately recorded as a "total loss," and in the nanosecond of that realization, I hit the water with a splash!

So to make matters worse, I was going to die! I found that out as soon as I realized that both of my legs were wrapped up in blackberries and my legs were held firmly about four feet apart, like a human slingshot. There I was; my upper torso treading ice water and the edge of the pond was directly under my knees. I could not turn over no matter how hard I wrenched; my arms and elbows were all that were keeping my upper body out of the water!

I made a giant effort to try to reach something behind me to try grasping anything so I could extricate myself enough to at least get my body on solid ground; figuring I could twist around to work on my thorns, but I kept having my only option of dog paddling to keep my face out of the water! The reality of dying in a few minutes didn't bring forth, "my entire life passing before my eyes" as they say; only sheer panic!

My mind appeared to have begun to be hallucinating then, as I had a mental picture of my prize elk coming out from behind the trees to watch me drown. It came right up to me as my face hit the water in absolute exhaustion, and I remember distinctly that my panic was all gone at that moment, and I had evidently resigned myself to going along with the inevitable; I was dying!

Then, suddenly, my mind was again functioning; I was on my back on the rocks and the blackberry vines were still wrapped tightly around my ankles; only I was flipped over 180 degrees from where I had been moments before! I rose on my elbows to see if it were real, and if the elk had somehow saved my life; and wiping my face clear, I saw a huge monster standing directly over me! It was a blurry mass of fur with huge legs and long, long arms. My mind was reeling as I knew that this was a Sasquatch! The mysterious animal of local legends had saved my life!

Then I had a coughing fit, spit up what seemed to be a few gallons of water; and then as my vision cleared, I saw the creature who had saved my life as it disappeared into the trees without even a glance back.

When I stumbled up to Hiram a couple of hours later, he took one look at the sling over my shoulder with the remnant of my rifle and scope hanging in pieces, and without saying a word, Hiram reached into his coat pocket and pulled out a flask of Johnny Walker. He put it in my hand as he helped me to a seat on a nearby log.

10 SQUELCHED BY SQUATCH

By Chris Nielsen ~ Bellingham, Washington

I was with a party of elk hunters near Kamloops, British Columbia. My father and three of his friends whom he had hunted with for many years had accepted me to accompany them to an area of British Columbia where they had hunted each year on the same property that was owned by one of Dad's business associates in Seattle.

Winter scene from Kamloops Lake, British Columbia

All five of us are from Bellingham, Washington, and the man whose land we would be on had been a "dream" of his; he had a fairly large log cabin built on it, a mountain stream flowed down past the cabin, and a short 50 yards below it, the stream formed a fairly good sized pond (about 100 feet across and maybe eight feet deep) before it flowed out the other end and continued downhill to a large river below. I will not mention the river, as the location would be a giveaway and the place's privacy would be compromised.

Dad had arranged by phone to have a guide, horses and mules, and a fully stocked supply of food and other necessary extras to make it as enjoyable as possible.

Each year, this party had hunted the same area, and each year had been about the same, as they had always each shot an elk. The only difference was last year's hunt which I heard all of the details about on our first evening in the cabin. What was different about the previous year was what the oldest member of the group had run into the third day of the hunt; no one would tell me, just that I'd find out the next day!

Riley had decided that he would take it easy last year, so when they left camp to go their separate ways for the day's hunt, he chose an area that none of them had hunted before. Instead of descending deep down into the canyon and hunting the vast pine covered plateaus, Riley went due north of the cabin along a fairly flat area that wrapped around to the other side of the large monolith that acted as a landmark for these friends to always find their way back to their cabin among the large pines.

Since I had never hunted such rugged terrain, it was decided that I should accompany Riley; to which I readily agreed. He was a man of 65, and with him slowing down, and me barely able to keep up, we were a perfect match. Riley and I walked toward his destination; chatting mainly about what I should be learning; how to walk stealthily and quietly, where to place my feet, and how to walk almost like a "stick" from the waist up. The knack of proper stealth was to only move from the waist down and watch not to step on sticks, pinecones or loose rocks. After three hours of Riley's constant whispered instructions, I began to feel a confidence. Then, I tripped over a log and fell flat on my face on top of a bramble bush. So much for my "Davy Crockett act!" Riley was grinning ear to ear as he helped me check my rifle for damage, but the only marks remaining were on my pride.

Riley suddenly stopped after I had once again asked about his experience on last year's hunt that no one would tell me about. He pointed to a large, dead tree that had fallen so long ago that there were almost no branches remaining, so we stopped for lunch. We sat facing each other, both straddling the huge tree, and it was then that Riley told me about the "mountain ape" he met last year.

He thought the reason nobody wanted to talk about it was because in his excitement he hadn't told it right, because my dad and the others had looked at him like he was crazy and they kept poking fun by asking if he'd been drinking or maybe he'd fallen asleep and dreamed it. They gave him such a hard time about it that he gave up talking about it anymore. He told me he was glad to have my young eyes with him in case we ran across the mountain ape again. Then I learned that Riley had encountered a Sasquatch! He still didn't with to elaborate, other than to give me a very believable description of it.

After we had rested, we continued to walk close to the sparse vegetation that consisted of berry bushes, thick to thin brush, and smaller trees along the outside edge of the dense forest along this fairly flat area. There was a downhill slant that made us stick to the animal trails that would lead toward a dark, dense forested area beneath a high parapet up about a quarter mile ahead.

Off in the distance to our left the scenery was fantastic. There over the valley, were thin, white and gray, wispy clouds, with a myriad of mountain peaks protruding into the beautiful blue sky. It seemed like it went on forever to the fog shrouded mountains. We stayed at about the same elevation, so in case we managed to bag an elk, we could call for the main party to come with a mule to drag it back to camp.

All of a sudden, Riley told me to duck as he grabbed my arm; pulling me down to my knees, and I was about to ask why when I followed his gaze to heavily forested area directly ahead of the trail and about a 30 degree angle below us. I was expecting to see an elk, but instead, I was looking at a large creature that I could have easily mistaken for a gorilla, except for the fact that it moved more gracefully. It was screened by a heavy area of brush, and had we not been at such a steep angle, we would have missed it entirely.

It also seemed more like a man, as it was not stooped over like a gorilla, and it seemed shaped more like a large, hairy human than a hunchbacked ape. Its arms were longer in proportion to its body than a man, but when it moved, it was more like a human than the back and forth rocking of the body as the gorillas in the zoos act.

As we were both sitting on the ground behind a raspberry bush, the creature had not even seen us. Then we got our answer as to how we were able to get so close without it seeing us. The Sasquatch (we knew it had to be) was carefully inserting a long stick into a monstrously large cone that must have been a bees' nest. Then, it slowly pulled the stick out and licked the honey from the stick with its enormously long tongue. All the while it was swiping the air with its other long paw at the cloud of bees that were a constant swirling shadow of angry insects. As Riley and I both watched, not daring to move, the huge animal suddenly lifted its head and it looked like it was sniffing the air.

Then it focused its gaze on where we were squatting behind the brush, and within two seconds it was gone! We realized that it must have caught our scent, so we stood up to return to camp, as the way ahead dropped sharply downward and we didn't relish the thought of trying to pull an elk up out of that area even if we got one, especially since it was so late in the day. Returning to join our comrades, we met them as they also happened to be returning at the same time.

That evening at dinner everybody took turns discussing their hunts, but Riley and I kept quiet and just said we didn't see even a track.

Dad has always been able to read me and when he noticed my inquisitive glance at Riley, and then Riley giving a slight shake of his head, Dad demanded to know what was up. Riley gave an audible sigh and agreed to tell about our experience.

He began by reminding the others as to how much ribbing he took the previous year when he tried to tell his story. Then, he finished telling about our experience, to which they now paid close attention. Then he turned to me and gave me the chance to tell what I had experienced. When I finished telling what I had observed, I turned the floor back to Riley and he added his feelings on what his interpretation had been.

As a group, we elected to put our elk hunt on hold (we could always hunt elk) and we decided to return to where Riley and I had watched the Sasquatch that day and explore for this fascinating creature! All of us were familiar with Sasquatch stories, but up until now, we had all assumed there was an explanation to these imaginary sightings, but we were now a team of six believers.

The next day, we broke camp with an excitement that an elk hunting trip didn't have. Only two of us carried our rifles, but we all had handguns, which were always our constant companions anyway.

With Riley leading the way, we retraced our previous day's route. We certainly didn't expect to see a Bigfoot, but obviously we had to be very close to their home grounds.

When we arrived at the remains of the bees' nest it was deserted and in pretty rough shape with a few of the feisty critters making angry passes as us, but we spread out to look for tracks.

About 10 feet down the slope, one of our party crossed a small creek that had two different sized footprints in the soft clay/sand along the creek as it wound its way downward. The tiny stream split into many small rivulets and they in turn wound around through the heavy foliage that covered every area of this extremely wet climate. Everywhere, the steep slopes were almost totally covered with a deep, dark mattress of many varieties of plants ranging from deciduous trees to vines that entangle ones' feet and legs, making it almost impossible to traverse such terrain in any reasonable time.

Kamloops, British Columbia
By sundiver72 (Pixabay)

In addition to the tough going, here we were, standing out on this steep slope and conspicuous as a grizzly bear on a sandy beach! We quickly grew tired of letting ourselves down these slopes, and it was a genuine relief when we came to an old logging road. Its state of disuse was apparent by the fallen trees that crisscrossed the road, and in places, the frequent rains washed large parts of the slopes down to block the winding road. We were glad to be so slightly inconvenienced when comparing with what we had previously been through.

This road we were on was returning us in the same direction from where our main camp was, only about 1,000 feet downhill from the cabin, so we were getting closer to camp according to the familiar cliffs high above us. We came upon a deep ravine that seemed as though the ground had opened up, and we were now on a seemingly major animal trail. The five of us were walking single file on a two foot wide animal trail alongside the cliff to our left and two or three hundred foot drop into a jagged, blackish ravine on our right. The animal tracks varied from tiny squirrel tracks, bird scratchings, deer, elk and one big bear track. We found a rough plowed trail where it went upward from the main animal trail; we were glad for that, so we made our way up to what was a better pathway.

About another half mile and there I saw it! Across the chasm, walking with a slightly hunched gait was a dark gray creature that we all saw at the same time (you have all of our signatures on a separate sheet, swearing to what we saw!) The animal looked at us like it wasn't even worried about being observed.

As we watched it from across this deep chasm, it looked to be keeping an eye on us. The Sasquatch seemed to be busy with lifting something out of a shallow ditch that was a branch of another creek that came down from the cliffs above. The Bigfoot animal momentarily disappeared into this rocky gorge, and a couple of minutes later, a small whitetail deer came flying out of the hole followed by the Sasquatch. It must have discovered or chased the deer until it fell off the cliff, and that accounted from my Dad's comment (he had been ahead of us) when he said the creature was throwing large rocks down at something.

The large ape looked sternly at our party of smaller than itself intruders, then it casually tossed the deer onto its shoulder, turned and strode off into the widening valley ahead of it. It was totally out of sight in a few minutes, and it never even gave our party another glance. We stood there talking about what we had just witnessed, suddenly realizing in all of our excitement, not one of us even gave a thought to scrambling for a cellphone for a photo. Tyler said he thought of it, but was afraid to turn his phone on because the loud sign-on bell would scare it away when he hoped to watch it longer. That made sense to the rest of us, as cellphones were taboo on our hunts anyway.

There isn't more that I can recall or any of the rest of our party could add to this commentary, and our hunt ended four days later without even seeing a bull elk!

We have all composed and reviewed our story, and after returning to our homes, several of us have contacted the local authorities to make sure we hadn't lost our minds! The B.C. wildlife management services, that charges those monstrous licensing fees for us nonresidents, listened politely to the three of us that later visited their offices and told us they believed every word of our encounter. This was not the typical denial from a government office that we expected.

Then the minister of game management phoned his secretary to bring into the conference room the files on reported Sasquatch encounters and sightings. When two people came into the room with a large bin of files and placed it on the conference table, they opened it so we could see all of the Bigfoot reports. We thanked them and left for home. That sure felt like a wasted day; as it was about a 200 mile drive each way.

11 SASQUATCH IS WILD AND DANGEROUS

By Walter S. ~ Whitefish, Montana

I was raised in Whitefish, Montana where my great grandparents had homesteaded way back before there were even roads; more like trails. My parents had inherited the vast acreage from Dad's parents. My grandparents lived in the main house and Dad built us our home about a hundred yards away. He called the location "near enough for friendship and far enough from nosiness!"

Together, Dad and Granddad built a large herd of cattle and I grew up in a prospering family. I did all the things a ranch kid does, and even though it didn't lack for adventure, I wanted to see the world!

Whitefish, Montana
By WikiCapa at English Wikipedia (Transferred from en.wikipedia to Commons.) [Public domain], via Wikimedia Commons

During a tour in Vietnam, I spent a year learning how to "duck" and after I returned stateside, I ended up in West Virginia. That's where I got married and spent my career in the military.

When I retired after 28 years, I needed to get away from the constant swarm of people who began to seem to me as "sheep," just moving about and going nowhere. Our two kids were raised and gone, and one day I was talking to my dad and the conversation ended up with him asking me to be his partner on the ranch, since it was too much for him as he had run with a foreman and some ranch hands since Grandpa died. Now Mom wanted to travel some, so we agreed to try it and see how my wife would get along with the Montana winters.

So here we were, in the cattle business, but I still pretty much let the hands do most of the work as I wanted to assimilate over time, as a respectable retired person would!

Early the following summer, I treated my wife Jenny to her first camping trip. Saddling our favorite horses and with a mule carrying a weeks' worth of supplies and a tent, off we went to a high meadow where my dad and I, and my grandfathers before him, had regularly spent time hunting and relaxing when time permitted.

We left the ranch early, and with our faithful dog Dillinger, we kept up a fairly brisk pace, as we had miles to go, and I don't relish setting up camp after sunset. We made good time without mishap; the advantage of having horses and a dog smarter than I!

The old memories returned as we topped a short rise and began our descent along a narrow cattle and elk trail that steadily dropped gradually to a beautiful meadow that opened up just off the main trail into which we cut off to. We rode across the park-like grass to a tree sheltered area adjacent to a trickling stream. It was about two yards across and slowly headed into a deep and desolate canyon far off in the distance. Although I had been here often when I was growing up, my father and grandfather never ventured far from this spot, even when they were hunting elk and mulies.

A trail bridge over Bear Creek in northern Flathead National Forest
By Rmhermen (talk · contribs) [GFDL (http://www.gnu.org/copyleft/fdl.html) or CC-BY-SA-3.0 (http://creativecommons.org/licenses/by-sa/3.0/)], via Wikimedia Commons

Once I recall asking Dad if we could head further down the sloping valley where it appeared to widen, and one could see an opening off of the large valley that looked like many separate valleys that cut off in different directions, and the way he sternly answered my query made me never ask again. I have never forgotten the severe, almost frightened look on his face, and evidently he must have realized that he had been too blunt, as about an hour later when we stopped to rest in our hunt, Dad put his arm around my shoulders and apologized. He then told my why he wouldn't ever go out into that canyon area.

He told me that we he was about my age, his dad took him and his best friend on a hunting trip into that area. His buddy Dale and he were not yet old enough for hunting the big game, but they had their .22 rifles for squirrels and crows.

He went on to say that on their second day after setting up camp they were riding down into the large open area that I had wanted to venture into, and all of a sudden, a monster buck stepped out about a hundred yards away. They dismounted and Granddad hobbled their horses to allow them to graze in the clearing. He said it seemed like their trek would never end. They were trying to walk on the balls of their feet, because that's the way Granddad said the Indians walk, but it was very difficult, when suddenly Granddad dropped to his knees.

I remember at the point in Dad's telling me this story, he was visibly trembling as he continued, and I thought I could see tears beginning to form, as his eyes had the appearance of being elsewhere. He pointed ahead at a small mound on which there was a large, dead tree surrounded by rocks and brush that had been piled up; it was near where two fairly large streams came together and formed a good-sized river that I could hear crashing down through the mysterious dark canyon further beyond.

Dad went on to say that both Dale and he mirrored his father as they silently and slowly made their way closer to the rocky mound. They soon found themselves in tall grass beside the mound. Granddad told them to sit down and not move. He said they sat there silently for what seemed like a long time before they heard a sound that they couldn't identify. Dad's friend Dale was closest to where the noise had seemed to come from, and Dale slowly stood up until he could see over the tall grass. Dad went on to say, they suddenly heard a loud snarl and Dale jerked to the side as if to avoid being hit by something. Then came a loud scream that Dad said he would never forget as long as he lived! Quietly and tearfully he told me what happened next; his friend's head bounced against his leg. It lay there staring at him, and Dale's lips moved as if he was trying to say something and then his eyes closed!

Dad spent a long time sobbing on his knees after he finished his story, while I just knelt beside him, not knowing how to comfort him. He took a long time to recover enough to speak again. He finally was able to explain that his father heard the commotion and scream and came running. He said it took a long time before he could talk about this terrible experience, and Dad admitted that this was the first time since his friend's death that he had spoken about it. His own father went to his grave without ever again mentioning the incident! Then Dad explained that he thought he would be able to once again hunt this area, but he never came back again after that trip.

Dad told me a while later that both the physician and a local veterinarian had speculated that it may have been a pregnant female Sasquatch; due to when they examined the area, the smells, and analysis of some other signs on the trail that Dad didn't elaborate on.

So here I was with my wife, admiring the shadowy beauty of this deep, secretive canyon that seemed to branch in every direction. In trying to be as honest as I could, she sensed that I was holding back when she had wanted to go deeper into the maze of canyons confronting us, and I had abruptly stopped dead in my tracks!

I tried to make an excuse as to why I thought we should camp where we had stopped, but I knew that I was sounding confused and insincere, so I gave her the "Reader's Digest" version of Dad's story, leaving out the gory details, simply saying his friend got hurt and he felt the area was bad luck.

I had very honestly figured that I could handle the trip since it was my dad's experience, not mine, and I didn't think his telling me about the incident so many years before would still affect me, but I was just as unnerved as I was when I was a kid! Skipping completely by the truth of the incident, I just explained to my wife that Dad's boyhood friend had died in a freak accident, and I didn't feel we should go further. She took notice of the fact that I was trembling slightly, and I told her that Dad was a good storyteller.

I don't think she believed me, but she started making camp and we had a pleasant evening and spent two more days staying near camp and relaxing. I never saw any signs of strange animals or suspicious creatures, but for two nights, the evening winds brought us some very strange odors! I didn't say anything, but twice Jenny had remarked about smelling something like rotten meat.

On the way home, Jenny moved her horse alongside mine and said she had seen a bear in the distance before she caught back up to me. I thought that her horse had just had trouble on the last hill, but she said she purposely had stopped when the "bear" came out of a ravine and she watched him walk across a meadow. Jenny also mentioned that through her monocular, she plainly saw him and he had a large deer on his shoulder!

Now I got serious! Without telling her that bears don't walk on two legs and they never carry game on their shoulders; I explained what it was that she really saw. She and I became absolutely convinced that she had been watching a Sasquatch! What I wouldn't have given for that opportunity. I think I'll wait until my folks get back home and I'll have Jenny tell her story while I watch Dad's expression!

12 SASQUATCH IN BEAR COUNTRY

By Tom D. ~ Washington Co., Pennsylvania

My nephew Justin and I took a weekend trip to Emporium, Pennsylvania to do a bit of bear scouting. We arrived at noon and still had an hour trip up an old logging road, then down into a really deep valley along a very shady logging trail.

Cameron County Courthouse in Emporium, Pennsylvania
By Nicholas T (Flickr: Courthouse View) [CC BY 2.0
(https://creativecommons.org/licenses/by/2.0)], via Wikimedia Commons

We set up camp in the Clear Creek Driftwood area of the Allegheny National Forest. We prepared dinner, ate, and then went to bed in the back of the truck. At about 3:00 a.m., I heard something hit off the truck, and I dismissed it as an acorn until just before sunrise the truck shook and rocked like it was being pushed. We looked out, but didn't see anything.

Minister Creek, Alleghany National Forest
Alexvallejo [GFDL (http://www.gnu.org/copyleft/fdl.html), CC-BY-SA-3.0
(http://creativecommons.org/licenses/by-sa/3.0/) or GFDL
(http://www.gnu.org/copyleft/fdl.html)], via Wikimedia Commons

At sunup we grabbed our pistols and walked in separate directions along a native trout stream. I had gone for about 20 minutes or so when I got a really bad feeling; like I was being watched. I scanned the area with field glasses, but noticed nothing out of the ordinary. As I went on, I heard a limb snap. I was now on full alert! I scanned the area again, with glasses this time, and I spotted a huge patch of fur behind a large tree. I watched it; ready to dismiss it as a squirrel when a hand reached out and held onto the tree.

I took off for the truck, pistol in hand; the whole time, the snaps of limbs and huge foot falls shook my body; it was chasing me! I got back to truck, packed up on the fly and hit the horn. This was our agreed on signal for danger.

Ten minutes later here comes Justin on a full run. Behind him I caught three quick glimpses of two bipedal apes following him. He got in and we took off. He kept saying a bear hit him with a log; he had a cut on his head and was covered in dirt. I didn't tell him what I had seen.

Allegheny National Forest in September

In my hurry to get out of there, I lost the knife that my pap gave me. He has been with God for a while and the knife was important to me; so last weekend five friends and I went to look for it. When we got to the top of the valley logging road it was blocked by logs pushed over onto trees, but at the top of the pile was my gas can, cast iron skillet and Pap's knife. All the things I had left behind in my hurry to get out of the valley.

I'll never go back; it's their valley! I'll find other bears somewhere else.

Publishers note: We've heard stories before of Sasquatch building barricades as a warning to people to not come back to their territory. We are happy that Tom is going to heed the warning.

13 CHASED BY BIGFOOT

Michael S. ~State Road, North Carolina

Publishers note: This recount was written by Gary Swanson after several telephone conversations with Michael S.

Maybe looking back on these events and the sudden excitement in the family of Sasquatch that heretofore had lived in Hamptonville, North Carolina forests maybe was caused by the pending weather pattern. We know that animals have senses far beyond humans, and unfortunately there has never been the opportunity for any studies of certain species; especially the elusive Bigfoot!

This particular family, of somewhere between four and maybe more, had been living in a remote, mountainous valley where Michael and his wife had understandably become semi-familiar with their existence. All were comfortable with the arrangement until a month or so before the arrival of Hurricane Florence in 2018.

Michael had contacted us about a month before Florence devastated so much of the Carolinas, and we in afterthought, had begun to wonder if somehow these elusive animals had unknowingly been affected by some sort of internal barometer, such as the instinct that science says all species have that prepares them for pending seasonal changes well in advance of man becoming aware of exact variations in established weather patterns.

Of course all of this is pure conjecture; however we have seen many examples of this strange phenomenon since we became aware of these interesting beings that live among us, but whom we seldom see.

We received a call from Michael when he said he had heard of us, but before giving us his story, he first hoped to meet with a group who ran a regular series of Bigfoot trackers on television. They had expressed interest in interviewing Michael, but he said it seemed as though they had better stories to pursue, so he called us because his main concern was to somehow place his experiences on the record in case anything "happened to him," due to the increased aggressiveness of his furry neighbors!

Michael said that he had only seen these animals at distances, as whenever he approached, they would vacate the area far in advance of his arrival; however he saw enough of them to identify the group to be Bigfoot. Michael sent us about 20 photographs of the area, and many showed the broken branches he had described; however we had a sudden disconnect before the photos of the "teepees" of sticks that Michael described could be sent. Since the call was from a cellphone, we had to wait for him to try again.

A week after our first conversation, I received a call from Michael and he sounded as if he had been running. He said he was out of breath from walking at a very fast pace on his way back from a steep valley behind his home. He had made the hike back there to get some additional photos of those tepee-like structures, but he had suddenly been assaulted by heavy sticks and rocks that came flying from above, and with only short glimpses of his attackers, he had no time to snap a photo; only enough time to turn and run; he explained.

He did however; mention his two large dogs had really stirred up screeches, growls and apparent chest-thumping from the "mountain gorillas." Michael has a boxer and a pit bull. It seemed unusual that the boxer and the pit would both share the same personalities, but Michael said they both had the same temperament; tough!

Photo taken by Michael S.

I felt better hearing that he had his dogs with him, even though I could detect concern in his strained voice as he kept reporting his progress as he descended into the valley, and then the cellphone connection began to lose clarity and the signal finally broke entirely just as Michael began to talk about how they were closer to him than ever!

A bit later, Michael called back and his voice was clear now that he made it to a more open area, but his demeanor hadn't changed a whole lot. With relief in his voice, he started by saying, "I'm alive!" He said the Bigfoot pair that had been closest to him were following alone now, and the others, however many there were, seemed to have turned back. He suddenly said, "Listen, I'll hold my phone toward them;" I heard what he had described as a kind of growling and huffing and I asked if that could be the dogs, and he said, "No, those chickens are a block ahead. What you hear are the two Bigfoot that are closer to me now than before!"

What my ears picked up were a couple of sounds that reminded me of a heavy cigarette smoker trying to cough out a built up congestion (talk about a very scary growling!) Whatever it was, I remembered thinking how I wouldn't have wanted to be there without a gun! Michael asked me if I heard them, and then after I verified in the affirmative, he had caught up to the dogs and he said the two creatures had just turned back as his house had come into view across the immense field.

He was still close to a large, marshy area along a tree covered knoll and he expressed concern that the "Bigfoot pursuers" may be just on the other side; he believed they had turned back, but he said, "Maybe the big devils are on the other side of the knoll," and he began to run again. Just as he struggled to run and speak, his phone cut out again! Then when I tried calling him back, the line was dead. I tried repeatedly over the next few days to no avail! Now my mind was really roaming as I had not yet asked pertinent questions, which is our normal procedure to allow our submitters complete anonymity.

I spoke to the cellphone carrier, and they said they'd check it out, but the next day Hurricane Florence ended any chance of calling authorities or anyone else in the area. The only thing we could do was impatiently wait until the hurricane had exhausted its blustery show in this area. We were concerned for Michael's safety and it was a nervous time for us, because we were unable to contact him at all!

Two days later, Michael called. They had escaped damage and the Sasquatch had evidently been forced by the hurricane to abandon their secluded canyon, as Michael said he had hiked all the way through the canyon that he had been chased from previously.

There was no apparent sign of the Bigfoots' occupancy, and evidently a major wind from the slopes above had whipped through this deep gorge with such force that it had even knocked more trees down that further covered the somewhat brushy hideouts where Michael felt the creatures had lived.

Michael and I were in the midst of another conversation, and I was about to compliment him on his bravery, as I could hear the sounds of his labored breathing; which seemed like he once again was being pursued by the Bigfoot when the connection was lost again.

Publishers note: The interesting name of State Road is assumed to have been named when U.S. Highway 21 was first under construction and the locals took advantage of the flat, graded surface.

On Sundays it became a regular routine for the local residents to create a horseshoe pit on the road bed, and they'd spend the day pitching horseshoes.

Someone would say, "Let's go to State Road; so when it came time to name the town, it was already well established.

14 AFTER THE HURRICANE

Michael S. ~ State Road, North Carolina

Publishers note: This recount was written by Gary Swanson after several telephone conversations with Michael S.

Michael mentioned that these "Bigfoot characters" spent a lot of time in a large area of dead timber that had been destroyed by a forest fire that happened years ago, and so these critters would "whack" on these dead trees with large sticks and branches to create a horrible racket. Since he and his wife live between two short mountains, and with such a large un-forested area, the sounds would echo throughout the valley. The noisy beasts also had a habit of breaking branches off the dead snags and they even pushed over many 20 foot high trees.

This must have taken tremendous strength, because even though they were long dead, they still had some massive root structures connected to the forest of "gray ghosts" that became more spooky at dusk. To quote another party we spoke with who lived in the same neck of the woods, and whose roots went back to the fathers of the confederate states secession; Tom T. grew reminiscent when he said, "Like the gray ghosts of a forgotten confederate infantry regiment rising at dusk to stand guard over their fallen comrades."

This was the same area that Michael had been traversing during our first phone call and I distinctly remember, even with my poor hearing after a lifetime of handgun shooting; the snorts and a growl with a lot of huffing in the background as Michael held his cellphone back over his shoulder as he jogged his way toward the safety of his home!

At the end of one of our conversations, I had received a return call from a U.S. Forestry employee that I had spoken to about this incident, and this nice lady certainly sympathized with what I explained a friend of ours in her area was going through, and I told her that I couldn't and wouldn't disclose his identity. It was then that she swore me to the same pact as she said that she and her husband had been haunted by a similar group of Bigfoot that moved into an area behind their property, and this harassment went on for months until her husband and his two uncles took action. They went into the forest and the three of them proceeded to fire shotguns and revolvers in all directions (safely of course), shooting dead trees and raising "holy hell" for several hours all over the area. The lady said they must have scared the apes away as they never had another incident.

Until this last frightening episode, Michael said the only contact they had experienced was strictly the noise they heard at dusk and early mornings of the "godawful pounding" from the beasts rapping on the dead trees. This would often go on for hours!

Ironically, about a week after this conversation, we had a call from a vacationer who had experienced a sighting of what they said looked to be about a seven or eight foot ape-like animal, and it turned out to have happened about 10 or so miles from where Michael lives. The person didn't have any more information other than watching it cross the road, but she got our phone number from an Internet search of Sasquatch. She promised to get back to us if she has any further information.

Our Internet presence is where most of our submissions come from, and as with Michael, we love to hear from so many people and have a chance to share their stories while they retain their anonymity.

I heard from Michael again just prior to this book's release and he felt that the Sasquatch ranks may have thinned, because he hadn't had any recent incidents with their nightly "noise fests" of rapping on trees. He did say however, that the various "stick stacking" in the sort of tepee shapes had resumed.

We will remain in touch for any further developments!

Thanks again Michael!

15 THEY WERE SWINGIN'

Christopher Raimee ~ Maine

Deer season came a little early this year in Maine; at least the forests were still heavily covered with leaves and very few had even fallen when the season opened.

Winslow Homer ~ On the Trail ~ circa 1889

Not the best hunting for white-tail, but Dad and my two uncles and I went out back of our 180 acres of prime hunting grounds behind our family's property. We had three homes in these beautiful forests that our family had owned since shortly after the revolution. It was almost like our own small country, as our vast acreages were all adjacent to one another, and contiguous to my folks place and my two uncles' homes was a state forest. Talk about privacy!

Anyway, I felt quite honored to have been invited to join the brothers annual deer hunt; however, in my exuberance to join them, I omitted the final step of preparedness, and like a damn fool, I left my beautiful new L.L. Beans with laces dangling. Since they were leather and my khakis hung over them, no one else noticed either; that is until we hopped across one of our small creeks and I was suddenly lying face down in a freshly made set of deer tracks!

In retrospect, I would have liked to have had the presence of mind to quip something about checking the "sign," but I was in too much pain and too embarrassed. Anyway, nothing broken and once I was laced up and abused some by the others we again continued our trip.

About a half hour went by before my limping got worse, and so I wouldn't spoil the hunt for the others, we agreed that I would remain on stand at a point where two main hills rose on each side of a beautiful, but narrow meadow that meandered among a continuing series of very colorful maple and oak covered grassy knolls. Their customary path for the most success brought them all full circle to pass right by here on their return.

Finding a spot where I was able to climb about halfway up to a point on my chosen hill, where the trees were more sparsely growing and where I had a view of three other hills, I found I was directly opposite a long narrow trench; what I would call a sort of gorge. Trees on both sides of this wide channel were mostly oak with some white pine and a lot of Jack pine. I settled down with my back against a large oak, and there were small bushes dotting the hill at this elevation where I sat; I must have been invisible to the glance.

I guess I should preface my finding by allowing first that our family property was home to a family of Bigfoot! No, I'm not nuts; the Sasquatch is quite well known to live in our area and unlike most places where the animals are spotted and then harassed, no one in our family ever let the outside world know about them. Our properties were highly fenced by the roads and signed, so to keep outsiders where they would be considered to be "good" neighbors. Over my years growing up, I had often briefly seen these Bigfoot on family outings and picnics, and our families and their families kept our respectful distances.

Today, however, answered a lingering question that quite often came up at family gatherings, "Why did we not see more of their footprints?" We all had crossed areas where we would find a track or a series of their footprints, but as often as family members had seen the critters, why didn't we find more sign? Often, when just sitting around, this subject came up.

I sat there with nary a twitch, and suddenly, I had the answer! A large Bigfoot came into my vision on the right side of the gorge I was watching. It stood near the midpoint of a tall Norway pine and looked carefully around before it reached out and grasped a hanging limb on this giant pine, and then effortlessly swung out off the cliff and down to another pine several yards below, and in a continuous series of swinging from limb to limb, it swung downward and continued until it landed on the ground at the bottom of the gorge.

I watched as it then crossed over the creek running through the center of the valley, and without slowing to rest, the furry beast reached up, grabbed another tree branch and began swinging again from tree to tree; only this time he was traveling upward!

I must have sat there with my jaw agape, because I suddenly had a case of dry-mouth and I coughed slightly, hoping the big guy hadn't heard, but there he was, swinging steadily but in much shorter distances, but still angling upward.

I must have watched this magnificent display of incredible strength for at least 10 minutes. I had been absolutely enthralled with the

Bigfoot's magnificent performance! I was truly in awe of this creature.

The time passed more slowly after that, until I heard a whistle from one of our party and I stood stiffly. The movement shot a jolt of pain into my ankle that reminded me of my reason for sitting by myself all afternoon.

When everyone gathered, I was the only one of our party with a story to tell, and I took full advantage of the opportunity. I had everyone take seats on two ancient logs that lay parallel and only a few feet apart in this junction of valleys.

When I finished my time on stage, there were many open mouths, but a lot of skepticism until I met their challenge by pointing to several trees that I had seen the Bigfoot use in his travels. As was our family tradition to accept nothing without proof, I hobbled over to point out the several tree limbs I could remember seeing the big fellow use and my uncles and dad went over and up both sides to inspect the trees.

Sure enough, and thankfully for my sake, these animals must regularly use the same tree limbs for their continual travels, as everybody admitted that various branches on those particular trees had been worn smooth, while other trees alongside had rough bark as pine trees all do.

Not only had I proven my point, but on several occasions since that day, various relatives in our conclave had been making similar observations now that everyone knew where to look.

One of my aunts loudly expressed our common thought, "I wonder how many times I've taken a potty-break in those forests with those characters watching!"

16 DID SASQUATCH SURVIVE THE VOLCANO?

T. Timmerman ~ Cascade Locks, OR

I have waited many years to tell of this event, and I have often sat down to put it in writing, thinking no one would possibly believe me, so I always quit. Then, after I broke my leg a short time ago, part of the reading material my wife collected for my recovery included a couple of your "Sasquatch" books. After reading two and insisting my wife order the rest, I was amazed at how many others had similar experiences to mine with the Bigfoot animals.

So, after all of these years, since I now have the time, I'll recall as accurately as possible the scariest experience of my entire life! It all happened on Mt. Margaret in the Cascade Mountain Range. I was enjoying an entire year of unemployment after a large insurance settlement from an auto accident, and while I was recuperating; a lifelong friend who was also off for the next month, invited me on a fishing, camping and gold prospecting outing. I readily agreed since I was rehabilitated to the "hobbling" stage, and although I wasn't up to full strength, Bud said his Jeep would take us on a series of old logging roads that would get us to his, "favorite spot in the world!"

Soon after, I was finding out that there was a lot of truth to the phrase, "scared to death!" I won't describe our trip in to camp, because I try to block out that part. Let's just say, we made it in.

Mount Margaret was to the east of Mt. St. Helens. I say was, because when the Mt. St. Helens volcano erupted, she totally obliterated this almost 6,000 foot high mountain; it simply blew it completely away!

Today, I am likely one of the few people alive who can say I was there when it was a beautiful, scenic mountain.

Mount St. Helens, one day before the devastating eruption
By Harry Glicken, USGS/CVO (USGS photo of Mt. Saint Helens) [Public domain], via
Wikimedia Commons

Well, we set up camp and for the next three weeks, we both spent the most leisurely time of our lives. We fished, swam in in the horribly cold waters of Spirit Lake, and we spent hours digging, chopping and scratching with picks and shovels in the rocky creek bed for gold. We did find enough that we could say we had succeeded as "gold miners," but our entire collection of "color" was just that. Only enough that when piled all together it truly was gold in color! Maybe the old mining stories we had read were wishes?

Many of our days were spent in leisurely poking around the many trails in the area. There were lots of animal and human trails throughout the forests, but where our camp was, it was "slightly off the beaten path" as they say. It was just so inconvenient to get to it, and so difficult to even reach the lake, that we never even saw another human being the whole time we were there! We never even heard more signs of humans except for two gun shots far off in the distance. It was really secluded.

Spirit Lake
© *Steven Pavlov / http://commons.wikimedia.org/wiki/User:Senapa*

We had been there for about three days before we found out that we were not alone! Along about dusk, as we were enjoying our evening meal, an old, half rotten log suddenly came bouncing down the grassy hill just behind our camp. We heard it pounding on the ground above us as it lurched and glanced off the few boulders on the otherwise barren slope.

As we both stood and watched it come, we weren't too alarmed, as with each bounce, more of the ancient goliath flew off with every thump. By the time it finally came to a stop against a small, but tough group of saplings, there was little left of the once stately pine!

Then, as if an afterthought, Bud and I looked at each other and we both started to speak at the same time; "Did you see that?" We had both been so focused on the giant log's approach that it must have been our safety first concerns, but we later agreed that both of us had been acutely aware of a giant, ape-like creature that had been standing at the top of the hill, and we suddenly found ourselves silently, but attentively staring at the thick forest of timber atop the hill.

Buckling on our revolvers, we climbed hand over hand and bush by bush until we made it to the spot where the great tree had obviously been laying for many years. Both of us intently searched the long, dry grass around the last resting place of the giant tree for footprints of any type, but to no avail.

We next climbed up and around the giant stand of fir trees that reached into the azure blue sky above, and as I circled the giant on my right, I heard a snap of a tree limb off down below from where I stood.

Dimly, but very clearly, my eyes were meeting another pair of eyes down the slope about a hundred yards away. The other eyes were in the extra-large and very ape-like head that was atop an animal, and I immediately knew it had to be a Sasquatch!

I heard all my life about sightings, however this was different. The animal just stood beside the towering tree with its right arm extended, and its giant paw holding the tree; as a human would do. We must have stood motionless for over a minute, and I was still gasping slightly for breath from the exertion and the immense excitement while the Bigfoot acted as calm as can be. I swear you could hear my heart beating loudly, and I remember subconsciously wondering if the ape man could hear it also.

Next, as I was still visibly twitching to recover from my climb, the animal tilted his head slightly and I could detect a slight, but obvious wrinkling of its brow, while at the same time its left arm raised slightly as a human would do if they were about to ask a question! Its mouth opened slightly, while I was still almost in a trance while obviously trying my best to mimic a statue, and it seemed as if it was about to attempt to communicate with me, when suddenly, Bud sprang around a tree above and up the hill from me asking "See anything?"

That was that! The huge critter disappeared down the heavily forested slope and I never even heard the sound of its footsteps. The opportunity of a lifetime lost! The whole rest of the time we both watched and hoped, but nothing!

17 THIRTY YEARS OF SECRECY

Danny T. ~ Spokane, Washington

My Uncle Will recently passed away and I was helping Aunt Emma pick up the last few items from the old farm house they had lived in for over 40 years. She had sold the house and moved into town, as the 180 acres they owned wasn't farmed anymore and hadn't been for years.

Before leaving, Aunt Emma said she would like to see the "back 40" once more, and since I had my 4x4 pickup, I happily obliged her, as I knew how much they both enjoyed the remoteness of it all. Enough so, that her sale agreement with the new owner stipulated that crops could be grown on the land, but that the heavily forested part could not be cut or cleared.

The new owner had no problem agreeing to that clause, because he looked on all those trees as a privacy barrier that he planned to leave in its primitive state.

We went out to the main driveway and followed the county road the half mile to where a culvert allowed ingress to a very solid timbered gate; which I opened, and at Aunt Em's request, I closed behind us. She said it was merely a caution in case someone ignored the two large "no trespassing" metal signs on either side of the gate.

We drove downhill for over a quarter mile where we reached the tree-line and the overgrown pasture ended. Em asked me to drive a ways further into the forest and we soon came to a turnaround that was overgrown, but quite visible as it circled a gigantic, ancient oak tree, and then I parked so she could take a "last look" at what had been her home for so many years.

Then to my surprise, she got out, reached into the truck bed and pulled out the large bag of apples that she had removed from her pantry, and as she met my curious gaze, she said, "Back in a minute," which was my cue to shut up and wait; which I did.

About 10 minutes later I heard a loud, "Goodbye!" As she returned, I noticed that she had tears welling up in her eyes which she kept blotting with her hankie. When I asked if she was okay, she just waved me off politely, and I understood it was natural, but there was more that she eventually confessed to me about an hour later.

She said that she and Uncle Will, along with three adjoining property owners had been keeping a secret for 30 years. Em then explained that about 30 years prior, they and the neighbors had begun to miss a few things around harvest time. She and Uncle Will had once raised some sheep and a few cows and even a fairly large flock of chickens, and they, like their neighbors, began to lose an occasional animal now and then. She said at first, they figured it was hawks or one of the family of fox that they had seen, so they chalked it up to part of farming, as it really wasn't much of a problem.

Well, she went on to say they finally found out what was happening on a fluke; one moonlit night when of all things, Will came face to face with a Bigfoot!

He had heard the chickens squawking up a storm, so he sneaked out with his shotgun and crept around behind the barn, then circled back, and then when he heard another chicken squawk, he flipped on the remote for the yard light, and there was this giant, ape-looking thing staring back at him! Em said that Will told her, just as he started to pull the trigger, a "little foot" as he liked to say, reached out for the big one's hand and Will said it was like he almost killed the neighbor's kid! That was exactly how Aunt Em said he described it, and she said he was visibly shaken by the experience.

Well, I guess they and the neighbors got together and agreed to the most unusual pact ever conceived in the Spokane Valley. That was over 30 years of everyone planting a little bit more vegetables and raising a few more critters to support what Em called, "The Native People!" She fell silent after that, so I didn't press the issue.

I do plan, however, to one of these days invite Aunt Em to a Sunday dinner so my wife Penny and I can see if there is more to this story. If we can gather enough information, I'll contact you with a follow up to this story, and if Aunt Em agrees to share her information, I plan to ask her for an introduction to her neighbors as well. I'd sure like to hear it from their perspective!

18 PROVING PROFESSOR MELDRUM CORRECT

Chance ~ Northern California

Hi, my name is Chance and my friends, Wayne and Bob and I were gone; gone Squatchin' that is! It was a beautiful summer day in the mountains of Northern California. I was on a sabbatical from a highly intensive study that a group of us from college had been involved in.

Most recently, three of us in the group had attended a wonderful two day lecture by Doctor Jeffrey Meldrum; the well-known professor from Idaho. He had enthralled us with photographs, plaster casts and everything that seemed to me as absolute proof of the existence of Sasquatch! The three of us had for several years been seeking as much information as possible about this often seen and seldom photographed being that was so often seen living in our more remote forests.

Being as how all of us were fortunate to have parents supporting us entirely, looking back on it all we were basically at that time, a waste of skin! I hate to look back critically like this, but until Dr. Meldrum came along, I had zero interest in anything but being a goof off!

I'm certain my parents probably thought I'd never amount to anything, and that may still turn out to be true. It's not my fault, or my friends' fault that our parents are rich. I'll accept the curse of leisure without any apologies.

Anyway, the three of us friends now had a purpose that our common interest could focus on, and we packed every kind of camping supply the super-excited salesman at the outdoor shop could conceive of us ever needing. I think Dad's Suburban was packed with enough supplies for a platoon of explorers. Our folks seemed all excited to see us have a genuine interest in something besides golf and cruising!

So here we were in the beautiful forests of Northern California in the vicinity of an old gold mining region, and at the farthest back road/trail the 4x4 Suburban was able to take us. We had followed directions given to us by a friend of Professor Meldrum's to a place near Happy Camp, California; it was about as far as a miner's mule had ever gone!

As we unpacked the massive inventory, we heard ourselves speaking aloud that when we returned home later on, we felt it would be a blessing if we could just leave it all there, as we'd never be able to cram it all in like the salesman had.

We had a large tent for sleeping and a slightly smaller tent for storing some supplies and for our camp table and stools. The "burb" held our coolers so we could keep them plugged in to the accessory outlet to run things off the auxiliary battery. With sporadic runs to Happy Camp, we could stay the summer if we had to, in order to find Sasquatch.

In addition to Dr. Meldrum's undeniable proof, we had all read a myriad of stories by hikers, hunters, campers and gold miners about their encounters with the Bigfoot. It took us three full days to organize our camp, gather firewood, dig a large fire pit with no trees above it; and we cleared away all dried grass and bush, as these mountains would erupt in fire if they had half a chance. After all of that, we still didn't dare start a wood fire!

We were disciplined in our goal and we made it a point to keep the noise levels down. Even though we each carried a handgun, we didn't do any target shooting. We couldn't believe when we discussed it how well-mannered and serious we all were. This maybe was the first time we ever had a goal in life.

Each day two of us would strike out in a different direction, always leaving one at camp to guard our belongings and monitor the main two-way radio in case of emergency, but on the couple of occasions we experimented with it, communication was sketchy. We almost had to be making eye contact for our walkie-talkies to work.

On one trip, the trail appeared to curve around a large peak, and according to our trail map, it looked like it wound back in a complete circle to where we were standing. We decided to travel in opposite directions, and in case there was something being pushed by one of us, it would eventually be confronted by the other person halfway around.

We had made this decision because on three different occasions, we had caught glimpses of some sort of animals ahead of us, but never were able to catch up to one. We had a special interest, because on four occasions we had found footprints that could almost pass for bare human footprints, but they were really huge.

This day, Bob stayed in camp and Wayne and I walked to the large peak that we were most interested in, and when we reached the point where the trail split, we did also. It had rained slightly the night before, but the sun had already been out long enough to dry the ground sufficiently that if one kicked at the trail, dust would rise to object. We must have walked an hour, when I almost stepped on what appeared to be a human footprint; it was almost a perfect imprint of a bare foot. It did appear wider and longer than my size 11 ½ boot, but now, I wondered how could it just suddenly appear in the trail?

I backtracked a few feet and carefully examined an animal trail that led up to the flat butte about a 100 feet above me, but I knew that if someone or something jumped that far, the print would not possibly be so plain. I was just about to give up trying to solve my mystery when I thought to glance down and back; sure enough there was a fairly wide but steep trail leading downward to a much wider trail that went all the way along the cliff to a narrow, high walled slot canyon. Whatever it was had come up from there.

Mystery solved, I kept in pursuit of what I now considered to be a fresh track, as everything else, even an elk track, had been obliterated by last night's rain.

I followed the very plain footprints and picked up my pace to a fast walk; at times almost running, and soon, I was certain Wayne would be arriving from the other direction and then we'd see if there really was a Sasquatch or some drunken, naked goldminer! One would be as frightening as the other. I walked at this fast pace for about forty minutes more, and suddenly I heard a shout, but it was muffled by the wind.

I broke into a run, and I could see the end of the dirt wall just ahead of me, and on my right, I saw a lot of shorter buttes like the one on my left. I knew that in 100 feet, I was likely to meet Wayne, as I assumed that it was he who shouted. One, two, three more paces and suddenly, there was Wayne! We were both panting so hard we couldn't speak, but we just stood looking around with our hands, palms up, saying, "Where is it?"

Then we heard a loud sound of sliding dirt and rocks and we hurriedly ran to our left and down a steep slope, about six feet down; we half slid to the bottom and there we were on another ledge.

Below us, half running, half jumping down this long, steep slide was a light brown creature that looked exactly like the sketches and photos we had recently studied with Professor Meldrum! I wish the good doctor could be here now. Of course, we had no time to gather our thoughts or even stop shaking enough to retrieve a camera from our backpacks; only time to stare at America's mystery man as he stole a glance back to see if we were following him. Well, what a tale we'd have for Bob when we got back!

After the very long trip back, as we finally trudged the last few yards to our camp, we found Bob forlornly on a camp stool; the tents were both down and it looked like the Suburban had driven through them! Bob was drinking from a canteen, and before we asked what had happened, in his excitement, Wayne blurted out, "We saw a Sasquatch!"

Bob shrugged his shoulders and replied with something like, "Yeah, about 10 minutes ago he stopped in to say hello to me too!" Then Bob's sense of humor kicked in and he said, "I invited him to dinner, but he said he was mad at you guys for trying to chase him! I did offer him a raincheck, just in case." What a mess the big fellow had made!

We remained in the camp for another week, but in our hearts we knew that the big guy wanted no part of us.

Our parents half believed our adventure, but voiced a lot of skepticism about our lack of photographic proof. We decided to tear up our letter to the professor, because he's already been there and done that!

19 SASQUATCH IN THE "STATE OF JEFFERSON"

Sam and Traci ~ Redding, California

My wife and I live in Redding, California, and have been reticent about our experience on a trip to Oregon back in 2005. Due to my having been an employee of the city of Redding and my wife working for the school district, we didn't dare to join the frenzy that we heard often accompanies reports and sightings of Sasquatch.

Over our lives, we had enjoyed the reports of Bigfoot encounters and several of our friends had often told and retold of their adventures in the rugged mountains of Northern California and Southern Oregon.

Since inhabitants of both states that live in this vast, rugged area so often just call it "Jefferson State" for quick identification, this will pinpoint the area where our experience happened.

We had cut off Interstate 5 at Shasta and angled our way toward the coast on Hwy 299; we were headed for Hwy 101 and we planned on taking the coast road up to Gold Beach, Oregon. Plans changed however, when we hit Willow Creek and saw a rather intriguing sign directing us to Happy Camp to the north.

My wife had been a teacher and an avid researcher of the early gold mining history of these mountains, so her attention was arrested by the many accounts of miners that she had so often read aloud at the dinner table.

The decision was easy; we turned off toward Happy Camp and figured to drive through the mountains to Cave Junction, Oregon and then cut over to Hwy 101 and up to Gold Beach from Brookings, Oregon. After all, a vacation is a lot more fun if you have these spontaneous moments!

The drive was beautiful and with so many stops along the way to gather information and souvenirs. We were both rather surprised to see how many of these points of interest had things for sale such as books, dolls, figurines and even several masks that indicated a strong belief in the actual existence of Sasquatch.

At a gas station where we had filled up and purchased some snacks, there was a very spry old-timer sitting on a bench watching traffic go by, and as we were getting back in the car, he said, "Watch out for Bigfoot!" We laughed and returned his wave, and as we were picking up speed, we looked at each other and Traci wondered aloud about why the old-timer would have said that. We both thought it odd for him to have made that remark, because he couldn't have known that we would be aware of the existence of such a creature, but we shrugged it off.

We enjoyed the rugged mountain scenery, and the two lane highway wasn't bad like we anticipated, because there were numerous places to pull over on the shoulder to allow faster vehicles to pass. This allowed us to drive at a pace whereby we could enjoy the beautiful mountain surroundings without being rushed.

As we finally crested the top of a hill to the state line where California and Oregon meet there was a magnificent viewpoint so we just had to pull over. We parked and decided to stretch our legs, so I locked the car, and taking the camera, we both walked across a dirt crossroads that overlooked the beautiful valley from which we had just left.

As we came to the edge, we noticed a "Jeep road" that cut off of the main road and led into the forest. Then, this road cut off to the right and downward into the valley.

We noticed about six vehicles hastily parked in all directions on the flat area at the top, which was about 100 feet below where we now stood. As we edged over to observe, there was a whole crowd of people standing on a short ledge, and all staring off at the nearby steep hill leading up the opposing mountain.

As we were wondering what all the excitement was about, there were footsteps quickly approaching, and as we turned, we saw an Oregon State Police officer being led by an overly-excited civilian, and the man was excitedly chattering something about a "Bigfoot!" The man had a thick, foreign accent, and he must have flagged down the State Police when he saw the Sasquatch.

As the pair stopped only feet from us, the man pointed toward the hill where the small crowd was also looking, he said that his wife was the one in the yellow rain slicker at the front of the group.

Traci and I both turned interestedly and as we made eye contact with this man; he told us that a "huge Sasquatch" had run across the road just as they crested the hill, and several cars behind them had also seen it, and that accounted for all of the vehicles slightly below us. The trooper stayed calm, and it was obvious from his demeanor that this was not his first Sasquatch.

We had been trying to see what everyone was pointing at, but had not seen anything, when suddenly, the excited gentleman shouted, "Look over by the big white rock," and following his pointing forefinger, Trace saw it first, and unlike me, she had the presence of mind to take a photo with her cellphone while the ape-looking critter was just to the right of the large, light colored rock on the ridge! I know it's hard to see, but if you look really close, you can make out what could pass for an ape of some kind. She said it was perfectly clear when she snapped it, but then it was gone!

This animal though, didn't slouch over like a gorilla would. The only real similarity is in the fact that this animal seemed to have long hair, and large feet and hands (or paws). It was climbing rapidly over the ridge, and it was moving so fast, it looked more like a giant, hairy human, as we didn't detect any of the slouching or hunched-over ambling like apes do!

Every movement we could see was very man-like, whereas apes lurch from side to side when doing anything.

After our first spotting the creature, it remained well hidden by the trees that covered the steep sides of this ridge.

Periodically, someone would shout, "There it is!" Everyone would scramble for a better view, while the four of us had a much better chance to see it, even though we were much further away.

In the midst of the chorus of "oohs and ahhs," the poor animal missed several steps and often tumbled a few feet before it gained a purchase on the rocks again; and in discussing it later as we continued our drive, Trace and I both felt sorry for the poor beast that only wanted to be left alone!

We hope you can see the Sasquatch clear enough in the photo. This incident turned out to be the highlight of our vacation, and we told our story many times as we enjoyed relaxing at Gold Beach.

We have since visited your "Sasquatch Watch" Facebook page and read several of your books; which were quite enjoyable. That's how we found where to send this experience. We hope you print it in the future!

Publishers note: "Gold Beach" was so named because of the tremendous amount of gold mining all along the Illinois and Rogue rivers that carried enough gold in their fast moving water that the coastal sands were literally packed with gold!

20 I CHASED SASQUATCH ACROSS THE MISSISSIPPI

Bob D. ~ Duluth, Minnesota

Back in 1950, I was working on a restoration project to research and identify the survey markers placed by the United States Coast and Geodetic Survey about the turn of the century.

It was interesting work, because it took my partner Bill and I into some of the most wild and primitive places that neither of us knew still existed in the northern areas of the country.

This particular day, we were locating two survey markers in what was the Mississippi Headwaters State Forest up near Bemidjii, Minnesota. Not only is this the area where the river begins, but this locale is the home of the legendary Paul Bunyan as well!

I was on my knees, carefully digging with a small spade for the elusive section corner that the original survey notes indicated was there when I heard footsteps; assuming it was my partner, I reached over and lifted up his backpack with my left hand, while reaching with my right hand for mine, as it was lunch time.

When he didn't take his pack, the strain on my arm made me look over to see what he was waiting for. As I looked up, my eyes were looking into the huge, yellowish-gold eyes of a monster!

Headwaters of the Mississippi River
Randen Pederson [CC BY 2.0 (https://creativecommons.org/licenses/by/2.0)], via Wikimedia Commons

In my shock, I stumbled backwards and landed flat on my butt, and as I caught myself on my elbows, I stared up at this behemoth; looking down at me was a giant!

I still have a picture of it in my mind, and I'll never forget those moments as neither of us moved; my thoughts raced, as I have never seen anything like it before or since. I first thought it to be an escaped gorilla, but it didn't have the pug, squashed face; it was more like the face of a wolf, only with a much shorter snout and a wider and flatter face and nose.

Its mouth was open and I can't forget the very large, but short fangs; both upper and lower; the ears were like a wolf's, but a lot shorter. The fur on the animal was a mixture of colors from darker medium brown to an almost tan or blondish-gray around its shoulders and face, but everything is sort of blurry now, as I never wrote down the details.

Over the years, I have been able to piece all of my images together, but then at times, they fade again, but this encounter couldn't have lasted more than seconds, and I think every second must have turned into a minute in my memory!

As the huge animal and I were semi-frozen in time, I heard my partner let out a yelp, and maybe he saved my life; I don't know, but the Bigfoot was on a dead run off to the east and he leaped the deep creek about 10 feet before I got to it, and I was suddenly aware that my body was in hot pursuit of the ape man!

My body seemed to have acted without me knowing, and I do not know what my mind was doing; only I was following as fast as I could in hopes I might be able to see where it went. My efforts were soon thwarted as I realized that in all the excitement I had failed to leave behind my 25 pound backpack! That is, until I found myself falling headlong into the creek that the fleeing Bigfoot had effortlessly vaulted over!

When Bill caught up to me, I was busy brushing the water and wet grasses off of my clothes, and I must have made quite a spectacle, as Bill just stood and laughed and pointed at me, and laughed some more. Then, when he helped me up on the solid ground, he said, "Only my partner could fall into the Mississippi River where it was only three feet wide!" When I realized what Bill said was true, I couldn't help but laugh at myself.

When we reported in to our field office the next morning, the superintendent of our office was there to receive our report, but hardly in the manner I had anticipated. It turned out that there were numerous reports of similar encounters by other survey teams throughout the forests under our jurisdiction, and as with the others I was told, we were cautioned under strict orders to keep our experience quiet.

Our boss explained that with it being an election year and with "budget approval" number one on our department's list of importance, it would not do well for its agents to be seeing "Boogie men!" So we kept quiet about it, except for the fact that the nickname "Mississippi" hung over my head until I retired 18 years later.

Signpost marking the Mississippi River headwaters in Itasca State Park
Mark Evans from Orange City, USA [CC BY 2.0
(https://creativecommons.org/licenses/by/2.0)], via Wikimedia Commons

Publishers note: The Mississippi River gets its start in northern Minnesota in the small glacial waters of Lake Itasca. From there it flows some 2,320 miles to the Mississippi River Delta in the Gulf of Mexico.

Once, before I graduated from school, my father and I were hiking in an area close to where this incident occurred, and as I hopped over a small stream, Dad said, "Congratulations Son, you just walked across the Mississippi River!" There were no signs back in those days at all. It must have been prior to tourism as we know it today.

Gary L Swanson

21 TETON BIGFOOT

Jimmy P. ~ Hood River, Oregon

I developed a fascination with the Bigfoot creature after my grandmother invited me to spend a summer with her and Granddad at their summer retreat near Grand Teton National Park. They had a small vacation home in one of the most beautiful places I had ever seen! My grandparents loved spending time around the cabin and enjoying their neighbors, so I had plenty of time to myself.

The John Moulton Barn on Mormon Row at the base of the Grand Tetons, Wyoming
By Jon Sullivan, PD Photo. - PD Photo,
http://pdphoto.org/PictureDetail.php?mat=pdef&pg=8145, Public Domain,
https://commons.wikimedia.org/w/index.php?curid=3537847

Having been raised in Washington State, the forests were familiar, but it was such a pleasure to be able to walk in the woods that weren't constantly dripping water from the incessant rain.

Walking through the pine forests was really enjoyable and I would leave early in the morning with lunch and a canteen of water, oftentimes borrowing Grandfather's bicycle if my destination was a ways away.

I soon lost all fear of spending a day in the forests and as I practiced my navigation skills, I got better at returning to where I started. Once in a while whilst following some new adventure, I would find myself trailing a slow moving herd of bison as they grazed the open, grassy areas.

The grizzlies were not near this area, and I felt confident I could handle myself against a mad turkey or a pronghorn antelope, or one of the many red fox.

One morning I left very early, because I had twice seen some kind of animal I hadn't been able to identify in the two times I had caught glimpses of it. The first time I was coming down a well-worn animal trail down the side of a gentle hill, and up ahead, I saw what I thought to be a buffalo calf walking down a slope to a parallel trail, only a few feet below and running in the same direction I was headed, but much more hidden from view. When I got a chance, I double-timed ahead to a fairly clear area and crept stealthily through the brush which afforded me a view of the lower trail. The other route was at a much greater angle now, and at this point it was about 10 feet below where I now sat waiting for the buffalo calf to appear.

Being out of breath, I tried to slow my breathing, which to me sounded like a distant steam train. Ducking down the best I could, I awaited the bison calf when I heard gentle footpads oncoming below me and I hunkered down so I could barely see between two thick bushes alongside a large, sheltering balsam tree.

A few seconds later I got the shock of my life! There, only about 25 feet from me, walked, what at first glance I thought was someone wearing a costume, and just as I was about to call out a greeting, another being sprang into view with a sort of growl and grabbed the smaller one by the arm and jerked it off the trail and down the slope to the valley floor about a 100 feet away.

I stood up to get a better view of what looked surprisingly like two dark brown chimpanzees, and as I did, the big one caught my gaze, abruptly turned, and with the small one in tow, quickly disappeared into the heavy underbrush!

My vacation had just become more exciting! Over the next month, until it was time to head home and back to school, I saw these Sasquatch several more times, but always from a greater distance.

My grandparents were not at all surprised by my first and consequent sightings, and they accepted my stories with an, "Oh yes, they don't do any harm."

22 SERENADING SASQUATCH

Billy Cavalier ~ Truckee, Nevada

I am a member of a country western band which plays throughout the Nevada/Northern California circuit. We made an appearance in Oregon a while back and I picked up your book Hiking Sasquatch County at the Illinois Valley Visitors Center. Good read by the way!

What I want to mention is that it brought back memories of when I grew up in Cave Junction, Oregon. Not much of a town, and I see that it was even less than when I last went home to visit my cousins.

Back when I was growing up, I used to practice my guitar at a favorite little cove out on the Illinois River. The acoustics in this sheltered notch out of the cliff drowned out the sound of the river, but it made a super place for me to better hear my own music, as it seemed to surround me from all directions.

I went there as often as I could, because I was trying hard to get accepted into a band in the southern Oregon town of Medford. They met about a block from a house that was once owned by the movie actress Ginger Rogers. I mention that only because I know you once lived in that area.

Well anyway, as I practiced one day, a small rock fell off the cliff above me and plunked into the pond. I looked up in case it was a rockslide, but it was just the one rock. Then, a while later, the sun was hitting at a different angle and I could see the reflection of the cliff behind me in the pond, and I again noticed the small splash of a stone, and as I glanced at it, I became aware of the reflection of what I thought was a dog on the cliff edge above.

Since I was concentrating on a song, I just kept playing, but I kept glancing to see if the dog was still watching. That's when I realized that it wasn't a dog at all!

Whatever it was had evidently been kneeling, and then when I looked again, it looked like it was standing on two feet, and that's when I lost concentration on my music and I quickly looked up to see a Sasquatch! I knew right away what it was, because my dad and my uncle had almost shot one during the previous deer season.

I wasn't really all that shook up, because everyone in our town was well aware that the Sasquatch lived in these mountains and they were often seen catching fish along the Rogue and Illinois rivers. No one spoke much about it to outsiders though, because during one six month period there must have been a half dozen camera crews of

Bigfoot hunters hanging around the area, but they soon gave up when they received no help or cooperation from the local people.

I had many relatives who owned businesses here, and these morons were annoying everyone in town by their constant questioning, and after one of them made a statement in a local bar one night that they really hoped to "kill one of them for a million dollars in publicity," that ended their welcome in a hurry!

Everyone in town was against them and they began having flat tires, getting rocks thrown through windshields and all kinds of incidents. Within a week they were all gone and they never came back.

I spent many more afternoons practicing at my secluded and semi-private aquatic amphitheater impressing the resident Squatches, and I eventually landed a great gig with a Reno group that I have traveled with for 12 years now. Whenever I return home for a visit, I remember my first performances before a "live, but hairy" audience!

23 SWAMP MAN

John T. Shimmerhand ~ Biloxi, Mississippi

Until about two years ago I lived in New Brunswick, Canada for 14 years; and for the last two years I have relocated back to my birthplace in the good ol' USA!

Working for a pipeline company, I have traveled a lot until my recent retirement, and when I was hunting deer this last season with my cousin, we came across a set of tracks that were about the size of my size twelve boot, but they looked to be barefoot! The one print was pretty clear, because it was in an area of hardened mud and it had dried really plain and even showed the toes quite clearly.

At first, we both thought it was our nearby neighbor, ol' Charlie, 'cause he likes to keep his feet tough and he only wears boots when it's winter or when he goes to town. Looking closer at the tracks though and following them across the other side of the bog, we found a couple more and they were up on the far bank; and two prints were almost perfect impressions, like a plaster cast I remember making in school as a kid. I think my mother kept my "handprint" for years.

These two prints had hardened in the sun, but you could tell they weren't human, 'cause on both of 'em you could see claw marks! Very plain; there was one great big toe and four narrower and longer toes. The big toe prints had a sort of stub-like claw that was wide, but not all that long, and the other four toes had what looked like long, narrow claw-like extensions about two inches out from the ends; and at the very tips of the claws were deeper cuts that seemed almost like they dug in; maybe because the animal was climbing out of the low area and up onto the bank.

Neither Jimmy or I had stayed in the area after ending our schooling, but here we were some 40 years later, both within 10 miles of where we grew up, and after all these years we finally found sign of the swamp critter we had both heard about as kids, but never had before seen.

About the time we were getting ready to leave, there came ol' Charlie hiking across the ridgeline of his high ground, and when he saw us, he waved and came over to see what we were doing out in this "worthless ground" as he put it.

We retraced our steps to show Charlie the prints, and without a second's hesitation, he said, "Rougarou!" Whatever that meant. Then he explained that this was the same type of critter that many of the locals call the Bigfoot or Swamp man; pretty much one of a dozen such names. It seemed not too alarming since everybody just called it by what they wanted, as I remembered hearing long ago.

Charlie told us that his father had many stories that kept him in fear of this "Swamp giant" when he was growing up. He told us that his dad's brother, "Uncle Waldo," had lost his arm as a youngster when the Swampman had come out of the woods and attacked him when he and their mother were picking berries. Charlie said the Bigfoot had jerked the berry basket so violently out of Waldo's hand, that his little arm just snapped!

He said the local men took up arms and went on a week long hunt for the animal, and he couldn't remember quite well enough, but he thought the hunting party had killed two of them, but the common belief was that they had powers like witches! For that reason, his dad told him that his brother Waldo's arm was added to the bonfire where they burned the two Bigfoot creatures.

Ghastly as it sounds now, Charlie said back then they believed that everything that could have been contaminated or bewitched by these evil beings had to be totally destroyed. Charlie said his Uncle Waldo was never "quite right in the head" after that, and after all these years to see that the beasts may be moving back again had a lot of the old timers worried.

As he looked at the tracks we had found, Charlie let out a sort of sigh and said the town council needs to know about it. Then he turned to us and asked, "Ready to go huntin'?" We're still waiting for an invite.

24 SASQUATCH'S CAVE HOME

"Mysterious Bill" ~ Cave Junction, Oregon

For those people who are familiar with the Oregon Caves National Monument, you know there have been numerous sightings and encounters with the legendary Sasquatch that have been reported over the years. There are a great many mysteries connected with these areas between The Caves and the Selma, Oregon area. Word is, is that there is a network of caves that stretch maybe thirty-odd miles in a meandering line, roughly east to west.

My degree in geology was never all that important struggling in the South American oil fields and in the various other extreme climates I have been in, but now in retirement, I'm back in the place I can predict my own weather; Oregon, where if it's not raining, it's going to; real soon!

I have always heard about Sasquatch, and have many, many friends from childhood and more recent ones that have either seen or encountered the big animals!

Now, retiring back in Oregon, I recently purchased a home just out of Grants Pass. I bought a copy of "Hiking Sasquatch Country" for a guide to some of the trails I never had a chance to explore, because my parents moved away when I was a freshman in high school.

My wife and I have hiked the trails in the vicinity of the Oregon Caves and have toured The Caves. One item of interest that I saw on our guided tour was the fact that there were bones of an ancient grizzly bear in the cave and it's been ages since this species was first reported to be living in Oregon.

There are roughly 30 miles of "connected" tunnel networks associated with these caves and almost no further exploration has ever been done since the first discovery. The story goes that a man named Elijah Davidson and his dog "Bruno" were chasing a bear and it ran into the cave before Mr. Davidson could shoot it. That is the official discovery of the Oregon Caves.

There have been few lengthy explorations of this cave network over the years due to financial constraints, but even before moving back to Oregon, I wondered how vast this cave system might be. Currently, myself and three others are in the beginning process of trying to secure financial backing for our exploration of this cave system, but our interest would be to go in the obverse; beginning in the Marble Mountain area, which evidence indicates could be the terminus of this extensive system of cave network.

One minor exploration that I and one of my partners have conducted was on a parcel of land adjoining another parcel that is relatively close to a gravel pit on Marble Mountain. We went with the property owner to see a small cave way back on his property that he had known was there, but never found the time or had the desire to explore. Seeing how hard it was to get to, we could see why!

The three of us outfitted our party of amateur explorers by a trip to a mining supply store in Grants Pass; and we got just enough equipment to be dangerous. Then, instead of hunting for gold, we found bats! I won't go into where or when, but let's just say, "we picked and shoved" our way in this small opening until all three of us were able to squeeze inside!

After my claustrophobic heart beat slowed to slightly less than a hummingbird's, I was able to crawl through a small opening inside this narrow entrance and I got in far enough to be able to stand up straight; so great a feeling after being almost folded in half. My buddies waited in the first small foyer on the chance that this thing might cave in while I carefully moved around in my jagged environment.

I detected a rather repulsive smell that I find hard to describe even now. I can only liken it to an animal carcus I found on a hunting trip; it had been dead awhile and was infested with maggots and it reeked awfully. The smell in this cave reminded me of that.

Also, the floor of this cave was kind of sandy and there seemed to be a dusting of sand over all of the blackish lava rock that appeared to be the entire makeup of the rocky passage. The black seemed permeated with minute particles that sparkled in the reflection of my searching light beam.

Squeezing out of my oversized, heavy backpack, I was then able to climb up a few feet to where I had seen what appeared to be a large opening above some lava rocks that may once have been part of the ceiling. The rocks were really sharp, but I did finally manage to wedge my boot into a fissure in the rock to enable me to climb up to the large hole above the apparent cave-in area, and I pulled my pack up and looped it over a jagged rock corner.

I played my lantern over the other chamber and I could clearly see a very large cavern beyond me, and the cave's floor seemed to be covered with sand granules rather than the jagged lava. I could distinctly see some particles of fine sand fall from above as I hung there taking it all in.

This was fascinating to find such a large cave, and it was enough to convince me of the possibility that the geologists may very well be correct about the likelihood of a continuing system of connecting lava tubes that stretched the 30-odd miles to the Oregon Caves system!

As further fuel for my now burning fire of discovery, a small bat momentarily fluttered as it fled from its perch deeper into the darkness ahead to retreat from my interference with its slumber. I inched up further until I brought myself up to a sitting position on this sharp ledge; my head now almost touching the ceiling of this cave, perhaps 15 or so feet tall at the highest. Switching off my lantern, I brought to bear my high-powered beam so I could see how far ahead this cave stretched.

The difference was enormous! You'd have thought I had installed flourescent lights throughout what turned out to be a tremendous tunnel! Just as I was about to call to my anxious partners to "get in here," I was shocked to see a pair of eyes in the next cavern looking back at me! A few feet beyond the rubble pile in the large adit, about 20 feet ahead of me, were the fierce, yellow eyes of something larger than myself. If it wouldn't have been for the fact I was sort of squished against the sides and ceiling of this hole, I may have fallen inside. In retrospect; had I fallen in, I might not be reporting this now!

As it was, the other eyes and mine seemed to be locked together in our shock! Although it seemed like minutes, I'm sure it was only a matter of a couple of seconds before the animal (what I now swear was a Sasquatch) bolted further on into the seemingly endless passageway, and I could see it finally disappear in the gray-black cave, so far away that this place must me enormous!

Once my heart had dropped back out of my throat and I was able to speak, I remember yelling out to my friends, "There's a Sasquatch in here!" This is as far as I can report at this point, but I am so terribly excited about what this can mean; I just had to report this event! I hope to secure funding for a complete exploration.

Even though we are being blocked in our requests for exploration of this cave system at the current time, we are moving ahead with incorporating. Hopefully we will be issued some kind of legal documentation from the state, or most likely, the federal government, so we can legally procede with permits from the land owners and the other red tape. Then, I'm sure, we'll soon have good news to report.

I will stay in touch, as if we are successful in what I believe will be the biggest discovery in Oregon's history, you may with to write the entire story. Make plans to bring your cameras and join me and my partners in exploring Bigfoot's 30 mile long home!

Publishers note: We are quite pleased to receive this report on a subject that has been rumored for so long; that being, further exploration of the cave system that could turn out to be the longest in the country!

Thank you Bill, and please keep us informed on the status of your exploration and when the fundraiser begins so we may participate. We look forward to making the journey!

25 GRASSY KNOB BIGFOOT

"Oregon Joe" ~ Los Angeles, California

I finally got up the courage to confess my story that I have wanted to tell about for years, but I feared that telling anyone might cause my arrest, or at the very least, embarrass me terribly. Being able to have it documented under an alias in your book will be sufficient to at least record this strange experience.

I was in the southwest part of Oregon in the Grassy Knob Wilderness area. My buddy and I were staying at a campground near the Elk River with our wives.

Wild and Scenic Elk River, Rogue River-Siskiyou National
US Forest Service [Public domain], via Wikimedia Commons

We hadn't got together for years since my wife and I moved to Los Angeles, and he and his wife remained in Eugene, Oregon. This was our first reunion since we both retired from federal government employment. This is the reason I don't dare use our real identification, because it might mess up our retirement pay! I just feel the need to document our weird experience.

My friend, call him "Jack," and I had always enjoyed archery, and we brought along our bows and a whole bunch of second-hand arrows we had collected from swap meets over the years. You know the kind; those you could shoot at targets without bothering to even look for your arrow; kind of like plinking with a .22 rifle.

We had set up camp, and with our wives happy to sit in camp at a firepit area between our two trailers, Jack and I headed off with full quivers, food and water, and a compass. We were somewhere around where the Elk River began, as we found out afterward, but the only thing we could really go by was occasionally seeing the ocean.

We were doing our damnedest to outshoot each other by loosing arrows at rabbits, squirrels and trees, and then we stopped for a bite of lunch; and that's when it happened!

There, about 30 feet behind Jack stood a young deer. I guess it was a blacktail, but I'm no hunter. I remember whispering, "Don't move," while I carefully and slowly reached my bow and very quietly selected an arrow, over what felt like a half hour, but could only have been minutes, while Jack just sat stone still with a questioning look on his face as he watched me nock my arrow. He leaned slightly to his right as I took aim over his left shoulder and loosed my arrow.

At the twang, both Jack and the deer made eye contact; about the same instant as I made my best shot ever! The arrow hit the poor deer right under its neck and it dropped! I had never shot anything larger than a rabbit before, so I didn't know it would just fall over like that. That's when I felt guilty for what I had done; and obviously, here I am 14 years later still feeling guilty! This guilt, of course, was permanently etched into my subconscious by the verbal barrage of shame that my wife launched at me when I later told the story at camp.

Jack joined me at the now still body of the deer and it was heavier than I thought it would be for the size of it. We returned to our resting place to finish our lunch and so I could boast about my shot and try to justify my needlessly killing the poor animal. I even contemplated maybe skinning it, but I didn't have a clue as to how.

Grassy Knob Wilderness within the Rogue River-Siskiyou National Forest
US Forest Service [Public domain], via Wikimedia Commons

The answer to my dilemma came out of nowhere as loud thumps assaulted my eardrums and I envisioned a whole herd of its relatives trampling us to death, but the heavy pounding was all from one exceptionally large, furry beast that came thundering from the heavily forested area to my right. It grabbed the deer with one giant paw and bolted into a thick stand of nearby pine trees!

While I felt the blood coming back to my face, I knew instantly that this had been a Sasquatch like the ones Jack had been telling me about since we pulled up at his house two days before!

Jack said afterwards that this was the second time he had seen a Bigfoot, but the other one had been about a quarter mile away. This one left us both quite shaken, and we both sat there for quite a while exchanging, "Did 'ya sees?"

I guess my wife Sissy answered my guilty conscience best when she concluded that hopefully my deer would help feed a needy family of Sasquatch.

This very brief, but exciting event was the strangest experience I've ever had, and by the time we headed back home, I had learned so much more about the Sasquatch, plus actually experiencing one first hand. I'll never forget this "adventure of a lifetime!"

26 THEY GO ON ALL FOURS

Randy K. ~ Ketchum, Idaho

I guess my eyes have been fooling me, because I have evidently been watching a Sasquatch without knowing it!

I will soon be going into the U.S. Marine Corps., and my enlistment date was delayed a month so I could go directly to the school my recruiter had promised me. For the last few weeks I have gone around visiting friends and relatives for one last time before I take the final plunge.

I've been a guest at my uncle's house for three days now. Uncle Don injured his foot last month, so I've been kind of helping him and Aunt Rose around their place. They have a large piece of land near Picabo, Idaho, and it isn't used much anymore except for raising a small herd of sheep and a few ducks and chickens, but they love it and I could see why. It's beautiful!

Don and I have spent a lot of time on the large back porch, as it overlooks a small, spring-fed lake back behind the sheep corral. The porch reminds me of a western ranch house with its rustic log walls and matching half-log deck. It has high side rails, also a network of smaller, vertical slats that allow one to look out, but if you're 50 feet away from the deck, you can't see anyone sitting there.

That brings me to why I am reporting my experience to you now. For several days, Uncle Don and I have been watching a bear (so we thought), as it came out of the woods and walked through the field of tall grass and berry bushes as it made its way down to fish in the pond.

Don said the pond was connected to a wider series of larger ponds, and together they led to a beautiful lake where there were summer cabins around it.

Picabo, Idaho
Leaflet [Public domain], from Wikimedia Commons

The fish made it all the way up to my uncle's pond, and for nine months out of the year, there were a lot of sizeable trout. Uncle Don said jokingly that on his private lake there is no bag limit.

Anyway, with Uncle Don's bum foot and my lack of interest in fishing, the neighboring bear was having the place all to himself. This particular day, Don was chomping at the bit for some exercise, and he was healed enough to walk slowly with the assistance of two canes, so we carefully made our way down the "back way."

It was a fairly easy path and except for the series of rocks and flat stepping stones between short wooden steps between them, the grass alongside the path was over four feet high; couple that with the impenetrable series of raspberry bushes, I couldn't see anything. Even the house had disappeared except for the occasional glimpse of the tall light pole by the garage.

I let Uncle Don set the pace, and with periodic rests, we emerged suddenly on a flat, bare shelf that overlooked the entire valley. All of a sudden Don's hand was pushing me back as he himself ducked down, whispering, "There's our bear!"

I had forgotten all about the possibility of meeting up with a mean tempered bear, and my heart was pounding in my throat as I crept forward again on hands and knees. Then, as I could just barely see the bear, it reached out a long, hairy paw and plopped a large fish out of the water, but instead of scooping it out like I've seen bears do, it actually threw it over on a grassy spot alongside what looked to be several more fish.

Uncle Don looked over at me with a wrinkled brow and whispered, "That's not a bear!" Before I could ask him what it was, the huge animal must have caught our scent, because it stood up on its hind legs and gave out a loud snort! Then, it took of back up the hill on all fours. That's when I could see it was no bear!

It looked exactly like a bear when it raced up the hill, and its speed was incredible! As the Sasquatch reached the tall pines on top of the hill, it stood to its full height and just stared at us, and then it casually disappeared down the other side of the hill.

Uncle Don and I walked slowly back toward the house, and it was then that he admitted to the fact that he and Aunt Rose have been aware of the existence of these Bigfoot, and on many occasions, a family of them have helped themselves to a lamb or duck, but nothing excessive, and this only happens when we have a late spring or an early winter storm that interferes with their food supply.

Then I asked Don .if he knew that I was actually watching a Sasquatch, and he said he expected as much, because the bears have never come that close. He apologized for misleading me, but if I started talking "Sasquatch" instead of "bear," the whole country would descend on his property ruining the privacy for both them and their secret neighbors.

Their secret is safe with me, as I think if I started talking about Sasquatch, my Marine Corps. career may be jeopardized. Maybe I can someday discuss this experience openly; like when I make General!

27 BATS GUARD THE SASQUATCH

K. J. ~ Josephine County, Oregon

By the time I made the long hike, it was dusk and I planned to spend the night; I was about 500 feet from the entrance to our gold claim when I saw two bats shoot out of the adit and swerve up over me and down into the thick pines on the slope below. I had been aware of a bat occasionally being in the large area that had caved in long ago; leaving a high cavern in a side tunnel, so I wasn't surprised that it had found a mate, but for them to take flight before dark was not normal.

Of real concern to me was that the bats must have been disturbed by something inside the mine. Quite often in the past I had been in the adit and observed the resident bat hanging up in the larger cavern, but I never played my light directly on it, and I looked on the creature as our "watch bat!"

Now, as I approached within about 300 feet, my hand had reached down to unsnap the strap on my revolver's holster in anticipation of a trespasser. Bats are not normally disturbed by bears or other animals, and our claim is too remote for transients, so as unusual as it is these days; I suspected to maybe find a hiker seeking shelter or a recreational gold seeker trespassing purposely to find free gold!

My nerves were on edge, because this claim is quite remote, so playing my lamp around as I entered the adit, I turned quickly from left to right; looking first down the place where we had concentrated our most recent efforts and where my tools were. I had left them on a ledge where I worked during my last two-day visit. They were right where I had left them.

I hadn't gone but 20 feet when I heard the thump of footsteps, as whoever it was had scrambled from the side shaft and exited quickly, hitting the partial half of the ancient door that had once been used to protect the mine when we had equipment inside. (That was when gold was both more plentiful and more valuable!)

I had turned and bolted toward the door, and I made the distance in seconds; and without thinking, I shot through the entrance, not thinking of someone waiting around the side; but what I saw brought me up short!

There, about 60 feet in front of me was a huge, hairy Sasquatch! I knew immediately what it was, because there have been a couple of these big creatures in this area for years.

My wife and I have both seen them several times, and we never have tried to block our ancient mine adit because the only thing these visitors ever take is food, or their favorite thing of value is any kind of old tarp, or their very favorite prize are gunny sacks. We surmised they may use them for their beds; which we don't have a clue to where they are located, but most likely in an old, well-hidden mine shaft, of which these mountains are full of.

As usual with the few contacts we have had, and our family members before us, the Sasquatch do not appear all that afraid of us, as we never yell at them; that is, after our initial shouts of surprise. I noticed on those few occasions, that the Sasquatch also lets out a grunt of surprise, so it's obvious they don't like sharing their mountains with us either!

We, like so many of our fellow miners, accept these creatures as sort of our companions in a remote world of loneliness and solitude that goes with mining.

The stories of these mountain apes go back as far as there is any recorded history of mining or exploring these mysterious mountains. Back in the days of Indian wars, an entire army of Native Americans hid out in these mountains, so it's no wonder the Sasquatch can live here unmolested for the most part, because there are millions of acres with no roads!

28 EVER HEARD OF A BLONDE SASQUATCH?

Mary ~ Chehalis, Washington

> *Publishers note: We decided to include this recently arrived submission in this book at the last minute, because it is only the second report we have received thus far about the existence of a light-colored Sasquatch.*

I live in southwest Washington State, and both my husband and I have seen the Sasquatch several times in the last four years since we began thinning the brush on our rather secluded property. It's always been from a distance and never for more than a couple of minutes; mostly for mere seconds.

On this day, I was checking for fingerling trout that we suspicioned had been born in our shallow pond. I had followed the winding stream, walking gingerly on the tufts of matted grasses alongside when I saw movement in the deeper area. When in my kneeling position; my peripheral vision noticed movement further beyond the pond. There, beyond a couple of trees bent by the constant coastal winds was a moving patch of light tan hair.

My first instinct was that it had to be a deer, but coastal deer are normally black-tail and dark colored; from browns to almost black. I then thought it may be a whitetail from the other side of the coastal mountains, and then it turned and seemed much larger; so my mind now expected it was one of the neighbor's cows, even though I had never seen one in this area, and never in marsh grasses like we have around here.

Now I was really curious, so I rose up for a better view, and suddenly I was staring at a long haired and very light colored creature standing on two legs with the face of what had to be a Sasquatch!

Never having had but a distant glimpse of one before and never closer than a quarter mile, I was frozen in place! Here we stood, just curiously looking each other over, and I had a fleeting thought that I wondered if I looked "presentable!" Then, with my body frozen in place, in the depths of my mind, I wondered if this strange creature had similar concerns? How peculiar are the thoughts one has at times of extreme stress!

Before my thoughts could even organize, the strange animal was rapidly disappearing from view. It was racing across the swampy marsh as if it was on a solid path; feet hardly seeming to touch the soggy ground before springing forward again. It's silky, blondish-brown hair was flopping with every step like the long mane and tail of a palomino horse!

That was over a year ago, and neither my husband nor I have ever seen one of these Sasquatch again; and believe me, we've looked! It was really neat to have lived among them for a while! Especially after we realized that these creatures likely had lived here since long before we arrived. Our neighbors told us about many of their own experiences once we told about ours. It seems that nobody ever told us about them before, thinking we might feel that they were totally nuts; we probably would have at that!

ABOUT THE AUTHORS

Gary and Wendy Swanson lived in Grants Pass, Oregon for eight years, where they enjoyed hiking throughout the spring, summer and fall months with their dogs. In addition to their love of hiking, they also enjoy history. Southern Oregon is full of history of gold mining, logging and fishing along the wild and scenic Rogue River; so for them, it had been a great place to research history, explore the countryside and hike; all at the same time.

Although they have relocated to sunny Southern Utah, they still enjoy hiking with their miniature schnauzers, and as more Sasquatch stories keep arriving, they have begun to receive some very guarded and secretive information about an almost unbelievably evil creature called "Skinwalker." They are understandably leery of this research, because where Sasquatch is really interesting, the Skinwalker seems disgusting and dangerous. They received many stories of this very evil creature and published them in "Skinwalkers, Shapeshifters and Native American Curses" and "The Last Skinwalker."

If you have had a sighting or an encounter with Sasquatch, or even a Skinwalker, and would like your story published, the Swanson's welcome you to send your contact information, details of the encounter and any photos to swanliterary@gmail.com.